MNEMONICS AND PEARLS HANDBOOK

For Residents, Medical Students, Nurses and Pre-Hospital Personnel

MNEMONICS AND PEARLS HANDBOOK

For Residents, Medical Students, Nurses and Pre-Hospital Personnel

Steve C. Christos, DO, FACEP, FAAEM
Clinical Assistant Professor
Presence Resurrection Medical Center
Department of Emergency Medicine

William G. Gossman MD, FAAEM
Chairman, Department of Emergency Medicine
Creighton University Medical School

2016 Ed.

Mnemonics and Pearls Handbook For Residents, Medical Students and Pre-hospital Personnel
Copyright © 2016 by Dr. Steve C. Christos, FACEP, FAAEM and William G. Gossman, MD, FAAEM

All rights reserved. This book or any portion thereof may not be reproduced or used in any manner whatsoever without the express written permission of the publisher except for the use of brief quotations in a book review or scholarly journal.

First Printing:1999, 2001, 2004, 2007, 2010, 2011, 2012, 2013, 2014 Dr. Steve C. Christos, FACEP, FAAEM

ISBN 978-1-329-88634-6

Wenckebach Publishing
Chicago, IL

Layout and Cover Design - Sherry Gossman, RN, BSN
Cover Photo - Tony Macasaet, MD, FACEP

INTRODUCTION

MNEMONICS AND PEARLS HANDBOOK

Hello and thank you for having an interest in *Mnemonics and Pearls For Residents, Medical Students and Pre-hospital Personnel.*

This book contains mnemonics that will assist you in rapidly learning the essentials in medicine. The "pearls" will help you answer questions frequently asked in rounds or on board exams.

HOW TO USE THIS HANDBOOK
Each mnemonic is presented in the following format:

> **Causes of Chest Pain**
> **Mnemonic - (MAPLE)** 3 [1, with modifications]

- **Causes of Chest Pain** - This mnemonic covers the causes of chest pain
- **(MAPLE)** - the mnemonic
- 3 - indicates that each letter is used three times, for example the three "M's" for this mnemonic are:

> **M**yocaridal Infarction - **M**usculoskelatal - **M**yocarditis

- The superscript enclosed by brackets are references. In this example, reference 1 is Rogers PT: *The Medical Student's Guide to Board Scores*. Reference materials can be found on pages xx-xx. In this example, modifications were made to Dr. Roger's mnemonic. Therefore, the reference is listed as, [1, with modifications]
- If the reference/author is now known, ANK is placed next to the mnemonic.

Many of the "pearls" have been obtained from the references below:

- Rosen P, Barkin E, et al: *Emergency Medicine Concepts and Clinical Practice*. Chicago: Mosby Year Book. 5th edition 2002, 6th edition 2006, 7th edition 2010

- Tintinalli JE, et al: *Emergency Medicine: A Comprehensive Study Guide*, St. Louis: McGraw-Hill 5th edition. 2000, 6th edition 2004 and 7th edition 2011

Every attempt has been made to ensure accuracy.

If you have any suggestions, corrections, mnemonics or pearls you would like to share, please email: stevesfmc@gmail.com, billgossmanmd@gmail.com

INTRODUCTION

SPECIAL THANKS TO OUR FRIENDS AND REVIEWERS:

Karl Ambroz, MD, FACEP
Clinical Assistant Professor
Presence Resurrection Medical Center
Department of Emergency Medicine

Amy Archer, MD, FACEP
Attending Physician
Advocate Lutheran General Hospital
Department of Emergency Medicine

George Chiampas, DO, FACEP
Clinical Assistant Professor
Northwestern University Feinberg School of Medicine
Department of Emergency Medicine

Nicole Colucci, DO, FACEP, FAAP
Clinical Assistant Professor
Presence Resurrection Medical Center
Department of Emergency Medicine
Loyola University Medical Center
Department of Surgery, Section of Emergency Medicine

Brian Donahue, MD, FACEP
Clinical Assistant Professor
Presence Resurrection Medical Center
Department of Emergency Medicine

Matthew Jordan, MD, FACEP
Clinical Assistant Professor
Presence Resurrection Medical Center
Department of Emergency Medicine

William Parente, MD
Emergency Medicine Resident
Presence Resurrection Medical Center
Department of Emergency Medicine

Mark Postel, DO
Emergency Medicine Resident
Presence Resurrection Medical Center
Department of Emergency Medicine

Scott Plantz, MD, FAAEM
Associate Professor of Emergency Medicine
University of Louisville Emergency Medicine

Jason Langenfeld, MD, FACEP
Assistant Professor of Emergency Medicine
University of Nebraska Medical Center

Robert Oelhaf, MD, FAAEM
Attending Physician - Department of
Emergency Medicine
Penn Highland Healthcare,
Elk Regional Health Center, St. Marys, PA

Tony Macasaet, MD, FACEP
Medical Director, Emergency Services
Vernon Memorial Healthcare
Viroqua, WI

Special thanks to Kelly Delaney, Catherine Deregla, Denise Toriani, Sherry Gossman and Demetra for their efforts.

Good luck and we hope this handbook is helpful.

Steve C. Christos, DO, FACEP, FAAEM
Clinical Assistant Professor
Presence Resurrection Medical Center
Department of Emergency Medicine

William G. Gossman, MD, FAAEM
Chairman, Department of Emergency Medicine
Creighton University Medical School

MNEMONICS AND PEARLS HANDBOOK

TABLE OF CONTENTS

GENERAL 1
   HISTORY OF PRESENT ILLNESS 1
   DETERMINING THE ETIOLOGY OF DISEASE PROCESSES 1
   MORE HISTORY 2

CARDIOLOGY 3
   CAUSES OF CHEST PAIN 3
   SERUM MARKERS IN THE DIAGNOSIS OF ACUTE MYOCARDIAL INFARCTION ... 3
   TIMI RISK SCORE FOR UNSTABLE ANGINA / NON-ST ELEVATION MI 4
   TIMI RISK SCORE OF 30 DAY MORBIDITY / MORTALITY 4
   MAJOR RISK FACTORS ASSOCIATED WITH ISCHEMIC HEART DISEASE 4
   HEART SCORE FOR MAJOR CARDIAC EVENTS 4
   CAUSES OF ST SEGMENT ELEVATION 5
   EKG FINDINGS IN AMI 5
   QUICK EKG INTERPRETATION OF STEMI V1-V6 6
   AMI IN LBBB (SGARBOSSA CRITERIA) 6
   BRUGADA SYNDROME 6
   SGARBOSSA CRITERIA / LBBB PACED RHYTHM 6
   TREATMENT OPTIONS IN ACUTE CORONARY SYNDROME 7
   GLYCOPROTEIN IIB/IIIA INHIBITORS 9
   MANAGEMENT OF COCAINE-INDUCED ACS 9
   PERCUTANEOUS CORONARY INTERVENTION (PCI) - STEMI 9
   DIAGNOSTIC ANGIOGRAPHY IN NSTEMI PATIENTS 9
   THROMBOLYTIC AGENTS RECOMMENDED (AMI) 10
   ALTEPASE (TPA) AND HEPARIN DOSING IN AMI, CVA AND PULMONARY EMBOLISM ... 10
   CONTRAINDICATIONS AND CAUTIONS FOR FIBRINOLYSIS USE IN STEMI 11
   VENTRICULAR FIBRILLATION/PULSELESS VENTRICULAR TACHYCARDIA 11
   REVERSIBLE CAUSES OF PULSELESS V-FIB/V-TACH 12
   MONOPHASIC VS BIPHASIC DEFIBRILLATORS 12
   MEDICATIONS THAT CAN BE GIVEN DOWN ET TUBE 13
   ACC/AHA CLASSIFICATION OF RECOMMENDATIONS 13
   CAUSES OF ATRIAL FIBRILLATION 13
   ATRIAL FIBRILLATION 14
   APPROACH TO SELECTING DRUG THERAPY FOR VENTRICULAR RATE CONTROL ... 14
   THE ACC/AHA GUIDELINES FOR PHARMACOLOGIC CONVERSION OF 15
   WPW AND RAPID AF 15
   WPW = TACHYCARDIA WITH 16
   CHA_2DS_2-VASc SCORE 16
   ANTITHROMBOTIC THERAPY FOR ATRIAL FIBRILLATION 17
   MULTIFOCAL ATRIAL TACHYCARDIA (MAT) 17
   ATRIAL FLUTTER 17
   ANTIDYSRHYTHMIC DRUGS 18
   CAUSES OF PAROXYSMAL SUPRAVENTRICULAR TACHYCARDIA 18
   PSVT TREATMENT OPTIONS 19
   CAUSES OF PULSELESS ELECTRICAL ACTIVITY 19

TABLE OF CONTENTS

TREATMENT OPTIONS: ASYSTOLE / PEA 20
CAUSES OF PROLONGED QT ON EKG 20
CAUSES OF LOW VOLTAGE EKG 20
FOUR PHASES OF PERICARDITIS 21
CAUSES OF PERICARDITIS 21
PERICARDITIS PEARLS 22
CHF CAUSES / PRECIPITATIONS FACTORS / PEARLS 22
CHF ON CXR 23
TREATMENT OPTIONS IN CHF 23
ABDOMINAL AORTIC ANEURYSM (AAA) 24
AORTIC DISSECTION 24
HYPERTENSIVE EMERGENCY 25
HYPERTENSIVE EMERGENCIES – TREATMENT OPTIONS 25
CARIDOLOGY PEARLS 26
CARDIOLOGY PEARLS - INFECTIVE ENDOCARDITIS 27
CARDIOLOGY PEARLS - VALVULAR HEART DISEASE 29
CARDIOLOGY PEARLS 30

PULMONARY **33**
RISK FACTORS FOR DVT / PULMONARY EMBOLISM 33
VIRCHOW'S TRIAD 33
HYPERCOAGULABLE (THROMBOPHILIA) STATES 33
COAGULATION DISORDERS – INHERITABLE VS ACQUIRED 33
INCREASED ESTROGEN (CAUSES URINARY LOSS OF PROTEIN S AND ANTITHROMBIN) 33
D-DIMER 34
WELLS CLINICAL SCORE FOR DVT 34
D-DIMER + WELLS CLINICAL SCORE FOR DVT- DIAGNOSTIC STRATEGY 34
ANTICOAGULATION TREATMENT OPTIONS FOR VENOTHROMBOEMBOLISM (VTE) - DVT/PE 34
VTE PROPHYLAXIS 35
VTE TREATMENT 35
TREATMENT STRATEGIES FOR MAJOR BLEEDING FROM TSOACS 36
ANTICOAGULATION CASCADE 37
MECHANISM OF ANTICOAGULATION 37
WELL'S CRITERIA FOR ASSESSMENT OF PRETEST PROBABILITY FOR PE 38
PULMONARY EMBOLISM RULE-OUT CRITERIA (PERC) 38
PERC RULE 38
PULMONARY EMBOLISM - PEARLS 39
EKG FINDINGS IN PULMONARY EMBOLISM 39
PULMONARY EMBOLISM - PEARLS 39
PULMONARY EMBOLISM – PEARLS PREGNANCY 40
TREATMENT OPTIONS IN PULMONARY EMBOLISM 40
TREATMENT OPTIONS IF MAJOR BLEEDING STARTS AFTER THROMBOLYTIC THERAPY 41
TREATMENT OPTIONS IF MAJOR BLEEDING STARTS AFTER HEPARIN THERAPY 41
CAUSES OF DYSPNEA 41
CAUSES OF PLEURAL EFFUSIONS 42
CAUSES OF PLEURAL EFFUSIONS - SIDE EFFECTS OF DRUGS 42
PLEURAL EFFUSION PEARLS 43
CAUSES OF HEMOPTYSIS 43
COMMUNITY ACQUIRED PNEUMONIA (CAP) 44
MILD (AMBULATORY) CAP 44
COMMUNITY ACQUIRED PNEUMONIA (CAP) 44

TABLE OF CONTENTS

TREATMENT FOR NON-ICU – CAP **45**
TREATMENT FOR ICU – CAP **45**
SEVERITY OF ILLNESS SCORE **46**
IDSA / ATS GUIDELINES ICU ADMISSION DECISION **46**
EPIDEMIOLOGIC CONDITIONS AND RISK FACTORS RELATED TO SPECIFIC PATHOGENS IN CAP **47**
PULMONARY PEARLS **48**

RENAL-UROLOGY **51**
CAUSES OF HEMATURIA **51**
KIDNEY STONES **52**
NEPHRITIC SYNDROME **52**
TESTICULAR TORSION **53**
PENILE FRACTURE **53**
PRIAPISM **53**
UROLOGY PEARLS - TRAUMA **54**
UROLOGY PEARLS **55**
UROLOGY PEARLS – STDS **56**

TOXICOLOGY **59**
DIALYSIS CRITERIA **59**
INDICATIONS FOR DIALYSIS OF TOXINS **59**
OVERDOSES WHERE BICARB MAY BE A TREATMENT OPTION **60**
CHARCOAL INEFFECTIVE **60**
WHOLE BOWEL IRRIGATION **60**
RADIOPAQUE SUBSTANCES **61**
CAUSES OF NON-ANION GAP ACIDOSIS **61**
CAUSES OF ANION GAP ACIDOSIS **61**
METHANOL AND ETHYLENE GLYCOL **62**
SUBSTANCES CAUSING AN OSMOLAR GAP **63**
RESPIRATORY COMPENSATION FOR METABOLIC ACIDOSIS **63**
FACTORS INCREASING THEOPHYLLINE HALF-LIFE **63**
FACTORS DECREASING THEOPHYLLINE HALF-LIFE **64**
DRUGS THAT INCREASE DIGOXIN LEVELS **64**
POISONING ASSOCIATED WITH FEVER **64**
POISONING ASSOCIATED WITH HYPO-THERMIA **64**
DIAPHORETIC SKIN **65**
SIGNS AND SYMPTOMS OF CHOLINERGIC EXCESS **65**
TREATMENT CHOLINERGIC EXCESS **65**
SIGNS AND SYMPTOMS OF ANTI-CHOLINERGIC TOXICITY **66**
ANTI-CHOLINERGIC TOXICITY EXAMPLES **66**
SUBSTANCE CAUSING NYSTAGMUS **66**
MIOSIS **67**
MYDRIASIS **67**
CAUSES OF SEIZURES **67**
SEIZURE HISTORY **68**
MORE CAUSES OF SEIZURES **68**
CAUSES OF SEIZURES – WITHDRAWAL **68**
NON-CARDIOGENIC PULMONARY EDEMA **69**
LITHIUM TOXICITY AND TREATMENT **69**
5 STAGES OF IRON TOXICITY **70**

TABLE OF CONTENTS

IRON TOXICITY TREATMENT	70
TRICYCLIC ANTIDEPRESSANT (TCA) PATHOPHYSIOLOGY	71
CARBON MONOXIDE (CO) POISONING – PATHOPHYSIOLOGY	71
HYPERBARIC OXYGEN (HBO) DEFINITE INDICATIONS	72
ACETAMINOPHEN (N-ACETYL-PARA-AMINOPHENOL, OR APAP) OVERDOSE PEARLS	72
FOUR STAGES OF ACETAMINOPHEN POISONING	73
ACETAMINOPHEN OVERDOSE PROGNOSTIC INDICATORS	74
ACETAMINOPHEN OVERDOSE PEARLS	76
SALICYLATE OVERDOSE	76
SALICYLATE OVERDOSE PATHOPHYSIOLOGY	76
SALICYLATE OVERDOSE DONE NOMOGRAM	76
SALICYLATE OVERDOSE TREATMENT	76
METHEMOGLOBINEMIA	77
MUSHROOM POISONING	78
PEARLS - COCAINE, OPIATE, BARBITURATE, PCP, GHB, NMS, SYMPATHOMIMETICS	79
PEARLS - TODDLER TOX KILLERS	80
PEARLS - SULFONYLUREA OD, WITHDRAWAL SYNDROMES, CLONIDINE, LEAD, HEMLOCKS, DIG	80
PEARLS - BETA-BLOCKER, CCBS, SEIZURES, ANESTHETICS, STRYCHNINE, ODORS	81

NEUROLOGY 83

TRUE EMERGENT CAUSES OF SYNCOPE	83
CAUSES OF SYNCOPE	83
EVALUATION OF SYNCOPE	83
SAN FRANCISCO SYNCOPE RULE (SFSR)	84
CAUSES OF HEADACHE	84
HEADACHE PEARLS	84
TEMPORAL ARTERITIS (TA) / GIANT CELL ARTERITIS (GCA)	85
MENINGITIS PEARLS	85
STROKE SYNDROMES	86
BASILAR ARTERY OCCLUSION	87
CEREBELLAR INFARCTION	87
CVA PEARLS	87
STROKE MIMICS	87
STROKE CAUSES IN YOUNG PATIENTS	88
LACUNAR INFARCTS	88
ALTEPASE (tPA) IN CVA	88
tPA / HEPARIN	88
CEREBRAL PERFUSION PRESSURE	88
TIA	88
ABCD2	89
OCULOVESTIBULAR TESTING (COLD CALORICS) - DIRECTION OF FAST COMPONENT	89
OCULOCEPHALIC RESPONSE (DOLL'S EYES MANEUVER)	90
GLASGOW COMA SCORE	90
HORNER SYNDROME	90
ACUTE TRANSVERSE MYELITIS (ATM)	90
INCOMPLETE SPINAL CORD LESIONS	91
SPINAL CORD TRACTS	91
CAUDA EQUINA SYNDROME	91
COMPLETE SPINAL CORD SYNDROME ACUTE OR SUBACUTE	92
INSUFFICIENT EVIDENCE TO SUPPORT STEROIDS IN CORD INJURY	92
CRANIAL NERVES	92

TABLE OF CONTENTS

CORD INJURY SENSORY LEVELS	93
NERVE INJURIES	93
MEDIAN NERVE	93
ULNAR NERVE	93
RADIAL NERVE	94
LIFE THREATENING CAUSES OF ALTERED MENTAL STATUS	94
CAUSES OF COMA	94
ALTERED LEVEL OF CONSCIOUSNESS TREATMENT OPTIONS	95
CAUSES OF PERIPHERAL NEUROPATHY	95
NORMAL PRESSURE HYDROCEPHALUS - TRIAD	95
PSEUDOTUMOR CEREBRI (IDIOPATHIC INTRACRANIAL HYPERTENSION)	95
TABES DORSALIS	96
NEUROLOGY PEARLS	96

HEMATOLOGY / ONCOLOGY 99

CAUSES OF MICROCYTIC HYPOCHROMIC ANEMIAS	99
CAUSES OF THROMBOCYTOSIS	99
CAUSES OF EOSINOPHILIA	99
THROMBOTIC THROMBOCYTOPENIC PURPURA (TTP)	100
IDIOPATHIC THROMBOCYTOPENIC PURPURA (ITP)	100
HEMOLYTIC UREMIC SYNDROME (HUS)	100
METS TO BONE	101
HEME /ONC PEARLS	101
HEME PEARLS	102

PEDIATRICS 105

CONGENITAL HEART DISEASE - CYANOTIC	105
TETRALOGY OF FALLOT	105
MANAGEMENT OF HYPERCYANOTIC OR TET SPELL	106
CONGESTIVE HEART FAILURE	106
TREATMENT OF CHF	106
COMMON CAUSES OF NEONATAL SEPSIS / MENINGITIS (< 1 MONTH)	107
TREATMENT OF NEONATAL MENINGITIS	107
MENINGITIS > 1 MONTH	107
TREATMENT OF MENINGITIS > 1 MONTH TO 50 YEARS	107
TYPICAL CSF CHARACTERISTIC OF NORMAL & INFECTED HOSTS	108
FEBRILE SEIZURES	108
STREPTOCOCCUS IDENTIFICATION	109
STREPTOCOCCUS PNEUMONIAE "PNEUMOCOCCUS"	110
STREPTOCOCCUS PYOGENES	110
5 MAJOR MODIFIED JONES CRITERIA FOR RHEUMATIC FEVER	111
CAUSES OF MIGRATORY ARTHRITIS	112
SINUSES PRESENT AT BIRTH	112
CONGENITAL TOXOPLASMOSIS (TOXOPLASMA GONDII)	112
DIAGNOSTIC CRITERIA FOR KAWASAKI SYNDROME	112
REYE SYNDROME	113
DIPHTHERIA	113
CHILD ABUSE	113
HENOCH-SCHÖNLEIN PURPURA PEARLS	114
MECKEL'S DIVERTICULUM – RULE OF 2'S	114
RECTAL PROLAPSE (PROCIDENTIA)	114

TABLE OF CONTENTS

GASTROSCHISIS AND OMPHALOCELE **114**
ANAL PRURITUS **115**
HIRSCHSPRUNG'S DISEASE **115**
INTUSSUSCEPTION **115**
MALROTATION **115**
HYPERTROPHIC PYLORIC STENOSIS **116**
PEDIATRIC PEARLS **116**
APGAR SCORE **119**

SURGERY / GI / TRAUMA **121**
SMALL BOWEL OBSTRUCTION - HISTORY **121**
CAUSES OF SMALL BOWEL OBSTRUCTION **121**
SBO PEARLS **121**
CAUSES OF LARGE BOWEL OBSTRUCTION **121**
CECAL VOLVULUS **122**
CAUSES OF ILEUS **122**
APPENDICITIS **123**
CAUSES OF PANCREATITIS **124**
DRUGS THAT CAUSE PANCREATITIS **124**
PANCREATITITIS - RANSON'S CRITERIA **125**
GALLSTONES **126**
CHOLECYSTITIS **126**
PORCELAIN GALLBLADDER **126**
HEPATITIS PEARLS **127**
HEPATITIS PROPHYLAXIS **127**
HEPATOTOXINS **128**
LIVER ABSCESS **128**
CHILD-PUGH CLASSIFICATION OF SEVERITY OF LIVER DISEASE **128**
CAUSES OF FECAL LEUKOCYTES **129**
NON-VIRAL DIARRHEA PEARLS **129**
DIARRHEA PEARLS **130**
CAUSES OF POST-OP FEVERS **131**
ARTERIAL OCCLUSION **131**
POST-OP COMPLICATIONS OF THYROIDECTOMY **131**
SURGERY / GI PEARLS **132**
TRAUMA PEARLS **134**
OB TRAUMA PEARLS **138**

ORTHOPEDIC **139**
DESCRIBING ORTHOPEDIC RADIOGRAPHS **139**
INTERPRETING C-SPINES **140**
HISTORY FOR C-SPINE **140**
UNSTABLE C-SPINE FRACTURES **141**
POSTERIOR HIP DISLOCATIONS **141**
GALEAZZI'S FRACTURE **141**
MONTEGGIA FRACTURE **141**
COLLES FRACTURE **142**
EVALUATION OF ELBOW RADIOGRAPHS IN KIDS **142**
THE OTTAWA ANKLE RULES **142**
BONES OF THE WRIST **143**
ROTATOR CUFF MUSCLES **143**

TABLE OF CONTENTS

FELTY'S SYNDROME	**143**
SYNOVIAL FLUID ANALYSIS	**144**
ORTHOPEDIC PEARLS	**144**

PSYCHOSOCIAL **149**
RISK FACTORS FOR SUICIDE	**149**
DEPRESSION	**149**
BORDERLINE PERSONALITY DISORDER (BPD)	**150**

OB-GYN **151**
ECTOPIC PREGNANCY (EP) PEARLS	**151**
RISK FACTORS FOR ECTOPIC PREGNANCY	**151**
ECTOPIC PEARLS	**152**
OB /GYN PEARLS	**152**

OPTHALMOLOGY **155**
ACUTE ANGLE-CLOSURE GLAUCOMA (AACG)	**155**
TREATMENT ACUTE ANGLE-CLOSURE GLAUCOMA (AACG)	**155**
HYPHEMA	**155**
CENTRAL RETINAL ARTERY OCCLUSION (CRAO)	**156**
CENTRAL RETINAL VEIN OCCLUSION (CRVO)	**156**
ACUTE VISUAL LOSS	**156**
OPTHALMOLOGY PEARLS	**156**

ELECTROLYTES **159**
CAUSES OF HYPERKALEMIA	**159**
RHABDOMYOLYSIS	**159**
TREATMENT OF HYPERKALEMIA	**160**
CAUSES OF HYPERCALCEMIA	**161**
PAGET DISEASE	**161**
SIGNS AND SYMPTOMS OF HYPERCALCEMIA	**162**
TREATMENT OF HYPERCALCEMIA	**162**
HYPOCALCEMIA CAUSES	**162**
HYPOCALCEMIA SIGNS / SYMPTOMS	**162**
HYPOCALCEMIA TREATMENT	**163**
ELECTROLYTE PEARLS	**163**
ELECTROLYTE PEARLS	**164**
CAUSES OF ANION GAP ACIDOSIS	**164**
CAUSES OF NON-ANION GAP ACIDOSIS	**164**

ENDOCRINE **165**
ADRENAL INSUFFICIENCY (AI)	**165**
NORMAL ADRENAL PHYSIOLOGY	**166**
TREATMENT OF THRYROID STORM	**166**
MYXEDEMA COMA	**167**
DIABETES INSIPIDUS	**167**
ANTERIOR PITUITARY HORMONE RELEASE	**168**
VITAMIN DEFICIENCIES	**168**

TABLE OF CONTENTS

    ENDOCRINE PEARLS 169

ENVIRONMENTAL 171
    HYPOTHERMIA, HYPERTHERMIA 171
    SPIDERS, SNAKES, BEES, SCORPION, SEA CREATURES 172
    ELECTRICAL, RADIATION, HAPE /HACE 173

DERMATOLOGY 175
    DERMATOLOGY PEARLS 175

INFECTIOUS DISEASE 177
    SYSTEMIC INFLAMMATORY RESPONSE SYNDROME (SIRS) 177
    HIV/AIDS 177
    LYME DISEASE 178
    BABESIOSIS 178
    RELAPSING FEVER 179
    EHRLICHOSIS 179
    Q-FEVER 179
    ROCKY MOUNTAIN SPOTTED FEVER 179
    TULAREMIA 179
    TRAVEL CHEMOPROPHYLAXIS 180
    DENGUE FEVER 180
    HANTAVIRUSES 180
    MALARIA 181
    LEPTOSPIROSIS 181
    LEISHMANIASIS 182
    HOT TUB FOLLICULITIS, FRESH/SEA WATER CELLULITIS, LUDWIG'S ANGINA, NEMATODES 183

MISCELLANEOUS PEARLS 185
    WMD PEARLS - DERM 185
    CUTANEOUS ANTHRAX 185
    BUBONIC PLAGUE 185
    ULCEROGLANDULAR TULAREMIA 185
    BLISTER AGENTS 185
    TRICHOTHECENE MYCOTOXINS ("YELLOW RAIN") 185
    SMALLPOX 186
    WMD PEARLS - PULMONARY 186
    INHALATIONAL ANTHRAX 186
    PNEUMONIC PLAGUE 186
    PULMONARY TULAREMIA 186
    WMD PEARLS 187
    BOTULISM 187
    CHOLERA 187
    RICIN 187
    BLOOD AGENT 187
    NERVE AGENTS 188
    CHOKING AGENTS 188
    RSI - PREMEDICATION 188
    ENDOTRACHEAL TUBE (ETT) SIZE 190
    PRE-INTUBATION ASSESSMENT FOR DIFFICULT AIRWAY   190
    POST-INTUBATION PROBLEMS   191

TABLE OF CONTENTS

KETAMINE PEARLS	191
PEDIATRIC AIRWAY PEARLS	191
PEDIATRIC TUBE SIZES	191
PEDIATRIC AVERAGE WEIGHTS AND ENDOTRACHEAL TUBE SIZES	192
DRUG INFUSIONS	192
LOCAL ANESTHESIA PEARLS	193
THE #20 - AN IMPORTANT NUMBER	194
POSTEROLATERAL - AN IMPORTANT WORD	194

REFERENCES **195**

GENERAL

MNEMONICS AND PEARLS HANDBOOK

HISTORY OF PRESENT ILLNESS

Mnemonic - (O P_3 Q R S_3 T)

O	**O**nset - What time did the symptoms start - What activity caused the symptoms
P3	**P**ain location **P**alliative (what makes the pain better) **P**rovocactive factors (what makes the pain worse)
Q	**Q**uality (sharp, dull, heavy, burning, squeezing, etc.
R	**R**adiation (arms, jay, back, groin, etc.
S3	**S**everity (use pain scale 1 to 10, ten being the most severe pain) **S**ymptoms associated (nausea, vomiting, diaphoresis, SOB, F/C) **S**imilar episodes in past
T	**T**iming (how long, constant vs. intermittent)

DETERMINING THE ETIOLOGY OF DISEASE PROCESSES

Mnemonic - (AN INDICATIVE DIFFERENTIAL DIAGNOSIS DD) ANK with Modifications

A	**A**llergy
N	**N**eoplasm

I	**I**nfection
N	**N**osocomial
D	**D**rugs
I	**I**ntoxication
C	**C**ongenital
A	**A**utoimmune
T	**T**rauma
I	**I**nflammation
V	**V**ascular
E	**E**ndocrine

D	**D**eficiency
D	**D**egenerative

If you come up empty with all the above consider PSYCH

GENERAL

MORE HISTORY

Mnemonic - (AMPLE)

A	**A**llergies
M	**M**eds
P	**P**revious medical history
L	**L**ast mealn/ **L**MP
E	**E**vents

CARDIOLOGY

CAUSES OF CHEST PAIN

Mnemonic - (MAPLE)3 (PCP$_2$) [1, with modifications]

M	**M**yocardial Infarction **M**usculoskelatal **M**yocarditis
A	**A**ortic dissection **A**ngina **A**nxiety
P	**P**E **P**neumothorax **P**neumonia
L	**L**ow H/H **L**ung CA **L**esions, Skin (Herpes Zoster)
E	**E**sophageal rupture **E**sophagitis G**E**RD

P	**P**yelonephritis
C	**C**holecystitis
P	**P**ancreatitis
P	**P**ericarditis

SERUM MARKERS IN THE DIAGNOSIS OF ACUTE MYOCARDIAL INFARCTION

Marker	Earliest Rise	Peak	Normalize
Myoglobin	1-2 hr	4-8 hr	1st Day
CK-MB	3-4 hr	10-24 hr	2nd Day
MB-Isoforms	2-4 hr	6-12 hr	1st Day
Troponin	2-4 hr	10-24 hr	5-12 Days

CARDIOLOGY

TIMI RISK SCORE FOR UNSTABLE ANGINA / NON-ST ELEVATION MI

Mnemonic - (AMERICA) [Emergency Medicine Journal 2008;25:122, provided by Dr. Jeff Kovar]

A	Age (> 65 years)
M	Markers (raised serum cardiac markers)
E	ECG (ST-segment depression at presentation; ST change > 0.5mm)
R	Risk factors (at least 3/5 for CAD -- DM, HTN, HL, Family Hx of CAD, smoking)
I	Ischemia (at least two anginal events in previous 24 hours)
C	Coronary stenosis (prior stenosis of 50% or more)
A	Aspirin (use in previous seven days)

TIMI RISK SCORE OF 30 DAY MORBIDITY / MORTALITY
[JAMA. 2000;284:835-842]

4.7% for a score of 0/1
8.3% for 2
13.2% for 3
19.9% for 4
26.2% for 5
40.9% for 6/7

MAJOR RISK FACTORS ASSOCIATED WITH ISCHEMIC HEART DISEASE

Age > 40
Male sex
Family history
Cigarette smoking
HTN
DM
Hypercholesterolemia
Obesity
Known CAD

HEART SCORE FOR MAJOR CARDIAC EVENTS
[Neth Heart J. 2008 Jun; 16(6): 191-196]

Predicts 6-week risk of major adverse cardiac event (MACE)

HISTORY	EKG	AGE
Highly suspicious +2	Significant ST-depression +2	≥ 65 +2
Moderately suspicious +1	Non specific repolarisation disturbance +1	45-65 +1
Slightly suspicious 0	Normal 0	≤ 45 0

RISK FACTORS		TROPONIN
Hypercholesterolemia, Hypertension, Diabetes, Smoking, + Fam Hx, Obesity	≥ 3 risk factors or history of atherosclerotic disease +2	≥ 3× normal limit +2
	1-2 risk factors +1	1-3× normal limit +1
	No risk factors known 0	≤ normal limit 0

CARDIOLOGY

The HEART Score is a propspectively studied scoring system to help emergency departments risk-stratifiy chest pain patients: who will have a MACE within in the next 6 weeks and who will not?
- It involves only a 1-time troponin, at admission.
- The rest of the score is based on age, history, risk factors, and EKG.
- Low risk patients have a score 0-3 and have a less than 2% risk of MACE at 6 weeks.
- MACE is defined as: all-cause mortality, myocardial infarction, or coronary revascularization.
- All other scores are high risk (risk increasing exponentially) and require further management and admission.

CAUSES OF ST SEGMENT ELEVATION
Mnemonic (ELEVATION) 2 [Acad Emerg Med, 1999;6:930, with modifications]

E	**E**lectrolyte abnormalities (↑ K+) **E**xcitation (WPW→ Delta wave)
L	**L**eft bundle branch block **L**eft ventricular hypertrophy
E	**E**arly repolarization **E**mbolism
V	**V**entricular paced rhythms **V**ariant angina (Prinzmetal's angina)
A	**A**neurysm (left ventricular) **A**MI
T	**T**rauma (contusion) **T**reatment (pericardiocentesis)
I	**I**ntracranial hemorrhage **I**nflammation (pericarditis/myocarditis)
O	**O**sborn J waves (hypothermia) The amplitude of the J-wave is proportional to the degree of hypothermia; does not relate to pH and is not prognostic [10, 5th ed., pg. 1981] **O**verdose-cocaine
N	**N**SSTT – wave change **N**onocclusive vasospasm - (Prinzmetal's angina, cocaine)

EKG FINDINGS IN AMI
EKG diagnosis of AMI = > 1 mm (0.1 mV) of STE in limb leads, and at least 2 mm elevation in the precordial leads

	EKG Findings	Coronary Artery Involvement
Septal	V1-V2	LCA → LAD → septal branch
Anterior	V3-V4	LCA → LAD → diagonal branch
Lateral	I, aVL plus V5, V6	LCA → Circumflex branch
Inferior wall	II, III, aVF	RCA (90%)
Posterior	V8-V9	LCA → Circumflex or RCA → PDA
	STD w/upright T V1	
	R/S ratio > 1 V1-V2	
Right Ventricle	V4R (II, III, aVF)	RCA → proximal branches

CARDIOLOGY

QUICK EKG INTERPRETATION OF STEMI V1-V6
Mnemonic - (SSAALL)

S	S	A	A	L	L
V1 **S**EPTAL	V2 **S**EPTAL	V3 **A**NTERIOR	V4 **A**NTERIOR	V5 **L**ATERAL	V6 **L**ATERAL

AMI IN LBBB (SGARBOSSA CRITERIA)

STE > 1mm concordant with QRS   5 points

ST depression > 1mm in V1, V2, V3   3 points (STD in V1-V3 concurrently = 3 points)

STE > 5mm discordant with QRS   2 points

Score > 3 suggest MI (90% specific, however 36% sensitive)

Clinical utility of criteria are insensitive and probably have relatively low utility

SGARBOSSA CRITERIA / LBBB PACED RHYTHM

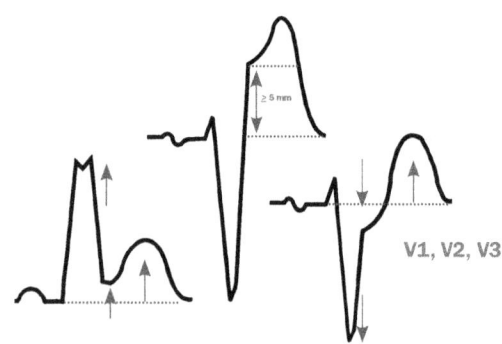

BRUGADA SYNDROME

Common cause of sudden death; genetic disease

ST segment elevation V1-V3 and/or "saddle deformity" of ST-T segment, with RBBB with or without the terminal S waves in the lateral leads that are associated with a typical RBBB

EKGs of Brugada Types 1-3 go to
http://lifeinthefastlane.com/ecg-library/brugada-syndrome

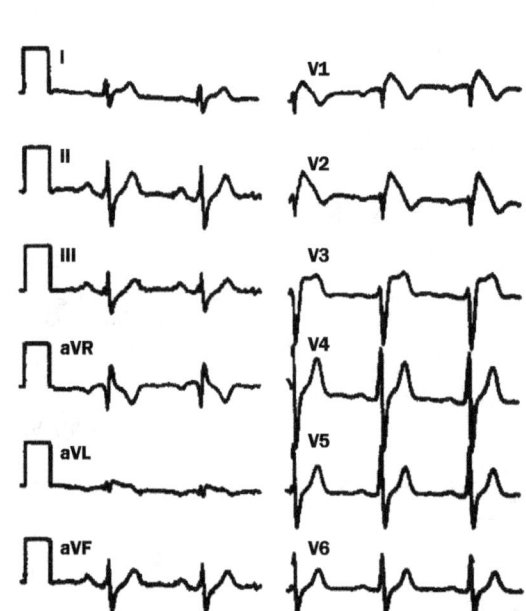

CARDIOLOGY

TREATMENT OPTIONS IN ACUTE CORONARY SYNDROME

Mnemonic - (HE BE MOAN) [21, with modifications]

[ACC/AHA 2007 Guidelines for the management of patients with unstable angina/non ST elevation MI J. Am. Coll. Cardiol. 2007;50;e1-e157; ACC/AHA Guidelines for the management of patients with ST elevation myocardial infarction. J Am Coll Cardiol 2009;54:2205-2241; 2010 AHA Guidelines for CPR and Emergency Cardiovascular Care; http://circ.ahajournals.org/cgi/content/full/122/18_suppl_3/S787 2015 AHA Guidelines Update for Cardiopulmonary Resuscitation and Emergency Cardiovascular Care; Part 9 ACS. Circulation. 2015;132[suppl 2]:S483-S500. DOI: 10.1161/CIR.0000000000000263]

HE	**He**parin (Class I, LOE A) **UFH** #1 Activates Antithrombin → inhibits activity of Factor Xa → Thrombin (IIa) formation from prothrombin inhibited #2 Activates Antithrombin → inactivates Thrombin(IIa) directly *See dosing to the right* **LMWHs** also activate antithrombin however: Primarily inhibit Factor Xa Minimal affect Thrombin (Factor IIa) *See dosing to the right*	**UNFRACTIONATED HEPARIN (UFH) IV** [Ref 29, 2010 AHA pg. 49] **STEMI and NSTEMI** • 60 U/kg IV bolus (max 4,000 U), • Followed with 12 U/kg/hr infusion (max 1,000 U/hr) **LOVENOX (LMWH) DOSAGES** [Ref 29, 2010 AHA pg. 50] **STEMI** - age < 75 years with normal CrCl • Fibrinolysis = 30 mg IV bolus • Follow with 1mg/kg SC q 12 hr (first dose 15 minutes after IV dose, with maximum of 100 mg for the first two doses only) • Dose for patients > 75 y/o = 0.75mg/kg SC q 12 hr without IV bolus (max 75 mg/dose for first 2 doses) • Dose for patients with CrCl < 30 mL/min = 1mg/kg SC q 24 hours • PCI = If last SC dose of lovenox was < 8 hours before • PCI → no additional dosing needed; if > 8 hrs bolus with = 0.3 mg/kg IV [www.druglib.com/druginfo/lovenox/indications_dosage] **LOVENOX (LMWH) DOSAGES** [Ref 29, 2010 AHA pg. 50] **NSTEMI** • Loading dose 30 mg IV bolus • Maintenance dose = 1mg/kg SC q 12 hours • Reduce dose if creatinine clearance < 30 mL/min = 1mg/kg SQ daily EMS: In systems in which UFH is currently administered in the prehospital setting for patients with suspected STEMI who are being transferred for PPCI, it is reasonable to consider prehospital administration of enoxaparin as an alternative to UFH. (Class IIa, LOE B-R) [*AHA, 2015, page 491]
BE	**BE**ta Blocker	**NSTEMI** Oral beta-blocker within 24 h (Class I, LOE A); Metoprolol 50 mg po Lopressor 5 mg IV at 5 min intervals x 3 = total 15 mg (Class IIa, LOE B) In the absence of contraindications
M	**M**orphine	NSTEMI patients: downgraded from Class 1, LOE C → IIa, LOE B potentially adverse effects of morphine in patients with UA/NSTEMI [*AHA, 2015] STEMI patients (Class I, LOE C) [Ref 29, 2010 AHA pg. 26]
O	**O**xygen	The usefulness of supplementary oxygen therapy has not been established in normoxic patients. In the prehospital, ED, and hospital settings, the withholding of supplementary oxygen therapy in normoxic patients with suspected or confirmed acute coronary syndrome may be considered. (Class IIb, LOE C-LD) [*AHA, 2015, page 491]

CARDIOLOGY

A	**A**ntiplatelet LOE = Level of Evidence Clopidogrel Limited evidence patients > 75 y/o	ASA 162-325 mg po/crushed/chewed; vomiting = 300mg rectal; ASAP If ASA contraindicated: clopidogrel (Plavix) 300 mg po [Ref 29, AHA pg. 39] **NSTEMI Conservative Therapy** (Class I, LOE A) ASA + add clopidogrel (Plavix) 300 mg po asap after admission **NSTEMI Invasive Strategy** (PCI within 4 to 24 hours) < 75 years old Give ASA + clopidogrel (Plavix) 300 to 600 mg or IV glycoprotein IIb/IIIa (GP IIb/IIIa) inhibitor **STEMI** < 75 years old ASA + add Prasugrel (Effient) 60 mg po prior to primary PCI (Class I, LOE B) *or* ASA + add clopidogrel (Plavix) 300 to 600 mg po before PCI or non-primary PCI [Ref 29, 2010 AHA pg. 39]; (Class I, LOE C) Fibrinolytic therapy: ASA + add clopidogrel 300 mg po (Class I, LOE B)
N	**N**itroglycerin (Class I, LOE B)	NTG 0.4 mg sublingual or spray every 5 minutes; if no improvement after 3 tablets/sprays, start IV NTG at 10mcg/min continuous infusion, ↑ 10 mcg/min every 3 to 5 min until relief or hypotension Hold if patients have recently taken ED meds sildenafil (Viagra) or vardenafil (Levitra), tadalafil (Cialis) Hold if extreme bradycardia (<50 bpm) or tachycardia (>100 bpm) in absence of heart failure Hold in patients with RV infarct [Ref 29, 2010 AHA pg. 54]

TREATMENT OPTIONS IN ACS

ASA alone relative reduction in cardiovascular mortality = 20 to 25 % [10, 5th ed., pg 1042]
ASA with lytics 42% reduction in mortality

ASA → Inhibits cyclooxygenase → ↓ thromboxane A2 production → ↓platelet aggregation and less arterial constriction

Clopidogrel (Plavix) and Prasugrel (Effient) → irreversibly inhibit ADP receptor on platelet cell membranes → block activation / transformation glycoprotein IIb/IIIa receptor

If CABG planned → hold Clopidogrel for 5 days, Prasugrel for 7 days [Ref 29, 2010 AHA pg. 39]

Prasugrel (Effient) is contraindicated in patients with history of TIA, CVA; use with caution in patients > 75 years old or < 60 kg due to risk of fatal bleeding, ICH and uncertain benefit; Not recommended in STEMI patients managed with fibrinolysis [Ref 29, 2010 AHA pg. 39, 40]

CARDIOLOGY

GLYCOPROTEIN IIB/IIIA INHIBITORS

→prevent fibrinogen/vWF crosslinking→↓platelet aggregation [Ref 29, 2010 AHA pg. 48]

Abciximab (ReoPro)
- STEMI with emergent PCI
    - 0.25 mg/kg IV bolus 10 to 60 minutes before PCI followed by
    - 0.125 mcg/kg/minute (max of 10 mcg/minute) IV infusion for 12 hours
- NSTEMI Invasive Strategy (PCI planned within 24 hours
    - 0.25 mg/kg IV bolus, then
    - 10 mcg per minute IV infusion for 18 to 24 hours, end 1 hour after PCI
- Must use with heparin

Eptifibatide (Integrilin) – NSTEMI Invasive Strategy
- 180 mcg/kg IV bolus over 2 minutes then begin 2 mcg/kg/min IV infusion repeat 180 mcg/kg IV bolus over 2 minutes in 10 minute
- Maximum dose (121-kg patient) for PCI: 22.6 mg bolus; 15 mg per hour infusion
- Infusion duration 18 to 24 hours after PCI
- Reduce infusion rate by 50% if CrCl < 50 mL/min

Tirofiban (Aggrastat) – NSTEMI Invasive Strategy
- 0.4 mcg/kg/min IV for 30 minutes and then continued at 0.1 mcg/kg/min IV infusion for 18 to 24 hours after PCI
- Reduce infusion rate by 50% if CrCl < 30 mL/min

MANAGEMENT OF COCAINE-INDUCED ACS
[http://content.onlinejacc.org/cgi/content/full/50/7/652 ; 28]

- SL NTG and a CCB (e.g., diltiazem 20 mg IV); avoid B-blockers
- If **ST-segment elevation** is present and the patient is unresponsive to initial treatment, immediate coronary angiography is preferred over fibrinolytic therapy.

- **UA/NSTEMI** → observed and managed medically for 9 to 24 h. If EKG and biomarkers are normal and the patient is stable, the patient can be discharged.

PERCUTANEOUS CORONARY INTERVENTION (PCI) - STEMI
- When performed within 90 minutes of patient arrival has been shown to be superior to fibrinolysis in combined end points of death, stroke, and reinfarction in many studies
- Primary PCI performed at a high-volume center within 90 minutes of first medical contact by an experienced operator that maintains an appropriate expert status (>200 PCI/year) is reasonable, as it improves morbidity and mortality as compared with immediate fibrinolysis (Class I, LOE A) [https://eccguidelines.heart.org/index.php/circulation/cpr-ecc-guidelines-2/part-9-acute-coronary-syndromes/]

DIAGNOSTIC ANGIOGRAPHY IN NSTEMI PATIENTS
- Persistent chest pain/symptoms/ischemia, heart failure, or arrhythmias, then diagnostic angiography should be performed (level of evidence: A)

CARDIOLOGY

THROMBOLYTIC AGENTS RECOMMENDED (AMI)
[Ref 29, 2010 AHA pg. 25, 46]

Presentation < 12 hours in context of signs and symptoms of AMI
- ST-segment elevation (> 1 mm in > 2 contiguous leads)
- Posterior-wall MI
- New or presumably new left bundle-branch block
- No exclusion criteria

Fibrinolytic therapy is generally not recommended for patients presenting between 12 and 24 hours after onset of symptoms based on the results of the LATE and EMERAS trials, unless continuing ischemic pain is present with continuing ST-segment elevation. (Class IIb, LOE B) [https://eccguidelines.heart.org/index.php/circulation/cpr-ecc-guidelines-2/part-9-acute-coronary-syndromes/]

Fibrinolytic therapy should not be administered to patients who present greater than 24 hours after the onset of symptoms.
(*Class III, LOE B) [https://eccguidelines.heart.org/index.php/circulation/cpr-ecc-guidelines-2/part-9-acute-coronary-syndromes/]

Altepase (tPA) – Accelerated infusion regimen is given over 1.5 hours
Give 15mg IV Bolus
Then 0.75 mg/kg (max 50 mg) over next 30 minutes
Then 0.50 mg/kg (max 35 mg) over next 60 minutes

Reteplase (Retavase)
10 U IV over 2 min
30 minutes later give second 10 U IV bolus over 2 minutes

Tenecteplase (TNKase)
Bolus: 30 to 50 mg, weight adjusted (not to exceed 50 mg)

ALTEPASE (TPA) AND HEPARIN DOSING IN AMI, CVA AND PULMONARY EMBOLISM

AMI
> **Altepase (tPA)** Accelerated infusion regimen is given over 1.5 hours [AHA, 2005 page 52]
> Give 15mg IV Bolus
> Then 0.75 mg/kg (max 50 mg) over next 30 minutes
> Then 0.50 mg/kg (max 35 mg) over next 60 minutes
>
> **Heparin** [AHA, 2005 page 55]
> Begin heparin with fibrin-specific lytics (tPA, Retavase and TNKase)
> Unfractionated Heparin (UFH) or
> Lovenox = ancillary therapy w/ fibrinolytic

CVA
> tPA dose = 0.9 mg/kg, max 90mg
> First 10% bolus over 1 min, remaining infused over next **60 minutes**
> Do not administer **heparin** or ASA during the first 24 hrs of fibrinolytic therapy [AHA, 2005 page 52]

Pulmonary Embolism = You will find three different protocols for tPA
- 100 mg over 2 hours (FDA approved regimen, most textbooks) *or*
- 15 mg bolus, then 85 mg continuous infusion over **2 hours** *or*
- Accelerated infusion regimen used in AMI

Hold **heparin** during fibrinolytic infusion. At the conclusion of alteplase infusion begin heparin infusion without a bolus when aPTT has decreased to less than < 80 seconds
[Circulation. 2005;112:e28-e32 Management of Massive Pulmonary Embolism]

CARDIOLOGY

CONTRAINDICATIONS AND CAUTIONS FOR FIBRINOLYSIS USE IN STEMI
[Circulation. 2004;110:588-636]

Absolute contraindications
- Any prior ICH
- Known structural cerebral vascular lesion (eg, AVM)
- Known malignant intracranial neoplasm (primary or metastatic)
- Ischemic stroke within 3 months EXCEPT acute ischemic stroke within 3 hours
- Suspected aortic dissection
- Active bleeding or bleeding diathesis (excluding menses)
- Significant closed head or facial trauma within 3 months

Relative contraindications
- History of chronic severe, poorly controlled hypertension
- Severe uncontrolled hypertension on presentation (SBP > 180 mm Hg or DBP > 110 mm Hg)
- History of prior ischemic stroke greater than 3 months, dementia, or known intracranial pathology not covered in contraindications
- Traumatic or prolonged (> 10 minutes) CPR or Major surgery (< 3 weeks)
- Recent internal bleeding (within 2 to 4 weeks)
- Noncompressible vascular punctures
- For streptokinase/anistreplase: prior exposure (> 5 days) or prior allergic reaction to these agents
- Pregnancy
- Active peptic ulcer
- Current use of anticoagulants: the higher the INR, the higher the risk of bleeding

VENTRICULAR FIBRILLATION/PULSELESS VENTRICULAR TACHYCARDIA
[2010 American Heart Association Guidelines for Cardiopulmonary Resuscitation and Emergency Cardiovascular Care]

- Start CPR, provide maximum oxygen, attach monitor/defibrillator →

- Unsynchronized Cardioversion = Biphasic preferred, manufacturer recommendation (120 to 200J); if unknown use max available; If only Monophasic available 360J →

- Resume CPR immediately – 2 minutes → check rhythm → shockable rhythm →

- 200J Biphasic or 360J Monophasic → Resume CPR immediately

- When IV available →
  Epinephrine 1 mg IVP repeat every 3 to 5 minutes

- CPR 2 minutes → check rhythm → shockable rhythm →

- 200J Biphasic or 360J Monophasic →
  Amiodarone 300 mg once IVP; second dose (if needed) 150mg IVP

- Extracorporeal membrane oxygenation: ECPR (or ECMO) techniques
  ECPR refers to venoarterial extracorporeal membrane oxygenation during cardiac arrest, including extracorporeal membrane oxygenation and cardiopulmonary bypass. These techniques require adequate

CARDIOLOGY

vascular access and specialized equipment. The use of ECPR may allow providers additional time to treat reversible underlying causes of cardiac arrest (eg, acute coronary artery occlusion, pulmonary embolism, refractory VF, profound hypothermia, cardiac injury, myocarditis, cardiomyopathy, congestive heart failure, drug intoxication etc) or serve as a bridge for left ventricular assist device implantation or cardiac transplantation.

There is insufficient evidence to recommend the routine use of ECPR for patients with cardiac arrest. In settings where it can be rapidly implemented, ECPR may be considered for select cardiac arrest patients for whom the suspected etiology of the cardiac arrest is potentially reversible during a limited period of mechanical cardiorespiratory support. (Class IIb, LOE C-LD)[https://eccguidelines.heart.org/index.php/circulation/cpr-ecc-guidelines-2/part-7-adult-advanced-cardiovascular-life-support/]

- Give 5 cycles of CPR → check rhythm → shockable rhythm → Start at top again

- Treat reversible causes

- CPR = Compressions, push hard > 2 inches (5cm), and fast > 100-120/min and allow complete chest recoil. Minimize interruptions in compressions.

- Rotate compressor every 2 minutes or sooner if fatigued. If no advanced airway 30:2 compression-ventilation ratio. If advanced airway – continue compressions and give 8 to 10 breaths/minute and check rhythm every 2 minutes. Avoid excessive ventilation. If PETCO2 < 10mmHg, attempt to improve CPR quality

- Narcan recommendation for agonal respirations superscript [AHA, 2015]

REVERSIBLE CAUSES OF PULSELESS V-FIB/V-TACH

- **H**ypovolemia
- **H**ypoxia
- **H**ydrogen ion (acidosis)
- **H**ypo / hyperkalemia
- **H**ypothermia
- **T**ension pneumothroax
- **T**amponade, cardiac
- **T**oxins
- **T**hrombosis, pulmonary
- **T**hrombosis, coronary

MONOPHASIC VS BIPHASIC DEFIBRILLATORS
[2010 American Heart Association Guidelines for Cardiopulmonary Resuscitation and Emergency Cardiovascular Care; S708]

- Monophasic waveforms deliver current of one polarity (i.e., direction of current flow)

- Biphasic is like an AC jolt, part of the shock goes from one paddle to the next and then reverses from the opposite paddle back. Less energy is required for the same effect.

CARDIOLOGY

MEDICATIONS THAT CAN BE GIVEN DOWN ET TUBE

Mnemonic (NAVEL) [ANK]

N	Narcan
A	Atropine
V	Valium
E	Epinephrine
L	Lidocaine

Also, isoproterenol [10, 5th ed., pg. 78]

ACC/AHA CLASSIFICATION OF RECOMMENDATIONS
[2011 American Heart Association; http://circ.ahajournals.org/manual/manual_IIstep6.xhtml]

Class I	Conditions for which there is evidence and/or general agreement that a given procedure or treatment is useful and effective
Class II	Conditions for which there is conflicting evidence and/or a divergence of opinion about the usefulness/efficacy of a procedure or treatment. IIa. Weight of evidence/opinion is in favor of usefulness/efficacy IIb. Usefulness/efficacy is less well established by evidence/opinion.
Class III	Conditions for which there is evidence and/or general agreement that the procedure/treatment is not useful/effective, and in some cases may be harmful.

Level of Evidence

Level of Evidence A	Data derived from multiple randomized clinical trials
Level of Evidence B	Data derived from a single randomized trial, or non-randomized studies
Level of Evidence C	Consensus opinion of experts

CAUSES OF ATRIAL FIBRILLATION

Mnemonic (ME WITH MITCH PhD) [Drs. Archer and Christos]
Atrial rate = 350-500, ventricular rate = 100-160; Irregular

| M | Mitral valve disease (MS, MR) |
| E | Electrolytes |

W	WPW
I	Intoxication/ETOH (holiday heart)
T	Thyrotoxicosis
H	HTN

M	Myocarditis
I	Idiopathic
T	Tox (CO, cocaine, amphetamines, heroin)
C	CAD
H	Hypoxia (COPD) *although more commonly Multifocal Atrial Tachycardia (MAT)*

P	PE
h	Hypothermia
D	Drugs (see tox above)

CARDIOLOGY

ATRIAL FIBRILLATION

If unstable (ie hypotensive) → "Synchronized" Cardioversion
Synchronized cardioversion is shock delivery that is timed (synchronized) with the QRS complex. This synchronization avoids shock delivery during the relative refractory portion of the cardiac cycle, when a shock could produce VF.

- Biphasic 120 to 200 J preferred, if not available then monophasic cardioversion at 200J [2010 American Heart Association Guidelines for Cardiopulmonary Resuscitation and Emergency Cardiovascular Care, S712]

- If the initial shock fails, providers should increase the dose in a stepwise fashion

- ↑ success of cardioversion if pretreatment with Amiodarone [JACC Vol. 48, No.4. 2006; August 15, 2006:e149-e246. Section 8.2.6.]

- Electrical cardioversion in Digoxin Toxicity → malignant ventricular arrhythmias

- Avoid electrical cardioversion if patient is on Digoxin unless condition is life-threatening, then use lower dose → 10 to 20 J [Ref 29, 2010 AHA pg. 43]

APPROACH TO SELECTING DRUG THERAPY FOR VENTRICULAR RATE CONTROL
[http://circ.ahajournals.org/content/early/2014/04/10/CIR.0000000000000041.full.pdf]

2014 AHA/ACC/HRS Atrial Fibrillation Guidelines[1]

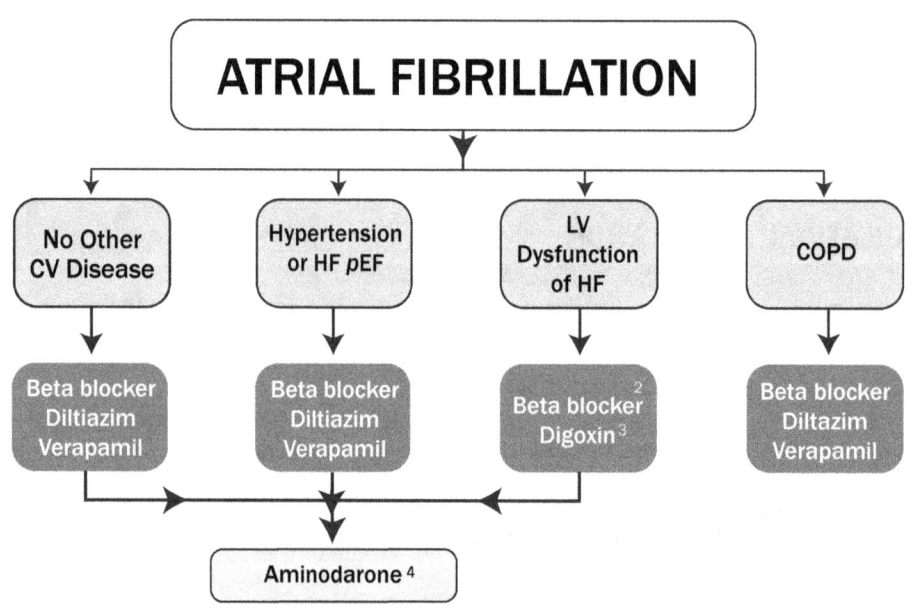

[1]Drugs are listed alphabetically.
[2]Beta blockers should be instituted following stabilization of patients with decompensated HF. The choice of beta blocker (cardio selective, etc.) depends on the patient's clinical condition.
[3]Digoxin is usually bot first line-therapy. It may be combined with a beta blocker and/or a nondihydropyridine calcium chanel blocker when ventricular rate control is insufficient and may be useful with patients in HF.
[4]In part, because of concerns over its side-effect profile, use of aminodarone for chronic control of ventricular rate should be reserved for patients who do not respond or are intolerant of beta blockers or nondihydropyridine calcium antagonists.

CARDIOLOGY

- Diltiazem (Cardizem 0.25 mg/kg IV bolus over 2 min; 15 min later if needed give 2nd bolus 0.35 mg/kg, then 5-15 mg/hr maintenance infusion

- B-blockers Metoprolol (Lopressor) 5 mg IV every 5" for total of 15 mg

- Amiodarone 150mg IV over 10 min, then 1 mg/min for 6 hrs, than 0.5 mg/min over 18 hours

- Digoxin 0.25 mg IV each 2 hours, up to 1.5mg
  1 hour delay before onset of action, peak effect does not develop for up to 6 hours

THE ACC/AHA GUIDELINES FOR PHARMACOLOGIC CONVERSION OF ATRIAL FIBRILLATION (AF)
[JACC Vol. 48, No.4. 2006; August 15, 2006:e149-e246. Section 8.1.5.]

Conversion of AF < 7 days	
Class I	Dofetilide (Tikosyn) oral Ibutilide (Corvert) IV Flecainide (Tambocor) oral or IV Propafenone (Rythmol) oral or IV Amiodarone (Cordarone) IIa oral or IV
Class IIb	(Less effective or incompletely studied agents) Disopyramide (Norpace) IV Procainamide (Pronestyl) IV Quinidine (Biquin) oral
Class III	Digoxin oral or IV Sotalol (Betapace) oral or IV
Conversion of AF > 7 days	
Class I	Dofetilide (Tikosyn) oral Ibutilide (Corvert) IV Amiodarone (Cordarone) IIa oral or IV
Class IIb	(Less effective or incompletely studied agents) Disopyramide (Norpace) IV Flecainide (Tambocor) oral or IV Procainamide (Pronestyl) IV Propafenone (Rythmol) oral or IV Quinidine (Biquin) oral
Class III	Digoxin oral or IV Sotalol (Betapace) oral or IV

WPW AND RAPID AF
- AVOID digoxin, CCBs, or beta-blockers!
- They may paradoxically accelerate the ventricular response → ventricular fibrillation. [JACC Vol. 48, No.4. 2006; pg e175] [10, 5th ed, pg. 1083]
- Consider Amiodarone or Procainamide (Class IIb)

WPW = TACHYCARDIA WITH

CARDIOLOGY

a. *Short* PR interval (less than 0.12 second)
b. QRS duration > 0.10s
c. The clinical hallmark = narrow complex supraventricular tachycardia at a rate of 150 to 300 beats/min
   [12, 5th ed, pgs 188-190]
d. Slurred upstroke to QRS complex ("delta wave")

WPW: PSVT in 40 to 80%, Atrial Fibrillation in 10 to 20% and Atrial Flutter in 5%

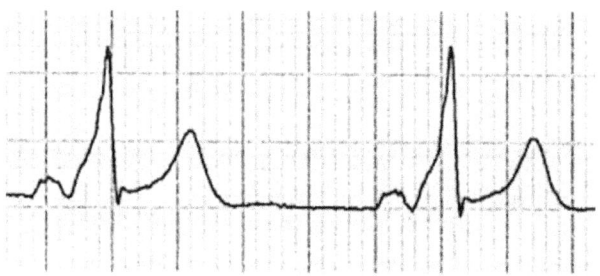

CHA$_2$DS$_2$-VASc SCORE
In patients with nonvalvular AF, the CHA$_2$DS$_2$-VASc score is recommended for assessment of stroke risk
[http://circ.ahajournals.org/content/early/2014/04/10/CIR.0000000000000041.full.pdf]

Definition and Scores for CHADS$_2$	
CHADS$_2$	**SCORE**
Congestive HF	1
Hypertension	1
Age greater than or = to 75	1
Diabetes Mellitus	1
Stroke / TIA / TE	2
Maximum Score	**6**

CHA$_2$DS$_2$-VASc acronym	SCORE
Congestive HF	1
Hypertension	1
Age greater than or = to 75	2
Diabetes Mellitus	1
Stroke / TIA / TE	2
Vascular disease (Prior MI, PAD, or aortic plaque)	1
Age 65 - 74 y	1
Sex category (ie female)	1
Maximum Score	**9**

Stroke Risk Stratification With the CHADS$_2$ and CHADS$_2$ and CHA$_2$DS$_2$ - VASc Scores	
CHADS$_2$ acronym[1]	Adjusted stroke rate(% per year)
0	1.9%
1	2.8%
2	4.0%
3	5.9%
4	8.5%
5	12.5%
6	18.2%
CHA$_2$DS$_2$-VASc acronym[2]	Adjusted stroke rate(% per year)
0	0%
1	1.3%
2	2.2%
3	3.2%
4	4.0%
5	6.7%
6	9.8%
7	9.6%
8	6.7%
9	15.20%

[1] These adjusted stroke rates are based on data for hospitalized patients with AF and were published in 2001 (201). Because stroke rates are decreasing, actual stroke rates in contemporary nonhospitalized cohorts might vary from these estimates.
[2] Adjusted stroke rate scores are based on data from Lip and colleagues (202). Actual rates of stroke in contemporary cohorts may vary from these estimates.

CARDIOLOGY

ANTITHROMBOTIC THERAPY FOR ATRIAL FIBRILLATION

- **AF and mechanical heart valves:** warfarin is recommended; target INR = 2.0 to 3.0 or 2.5 to 3.5 based on the type and location of the prosthesis

- **Nonvalvular AF, prior stroke, TIA or a CHA_2DS_2-VASc score > 2:** oral anticoagulants are recommended. Options include: warfarin (INR 2.0 to 3.0), dabigatran, rivaroxaban or apixaban

- **For patients with nonvalvular AF unable to maintain a therapeutic INR level with warfarin:** use of a direct thrombin or factor Xa inhibitor (dabigatran, rivaroxaban, or apixaban) is recommended.

- **Renal function** should be evaluated prior to initiation of direct thrombin or factor Xa inhibitors and should be re-evaluated when clinically indicated and at least annually

- **For patients with atrial flutter,** antithrombotic therapy is recommended according to the same risk profile used for AF.

- **CHA_2DS_2-VASc score of 1:** no antithrombotic therapy or treatment with an oral anticoagulant or aspirin may be considered.

MULTIFOCAL ATRIAL TACHYCARDIA (MAT)

1. > 3 differently shaped P waves
2. Varying PP, PR and RR intervals
3. Atrial rhythm between 100 and 180 [12]

- "Irregularly irregular" rhythm, "wandering pacemaker", "chaotic atrial rhythm"

- Most common cause of MAT = COPD. Other causes = CHF, sepsis and theophylline toxicity.

- Treatment of MAT is directed toward underlying disorder

ATRIAL FLUTTER

Atrial Rate = 250 to 350 beats/min

The 2:1 conduction ratio accounts for the classic (although not exclusive) EKG appearance of atrial flutter as a **narrow complex tachycardia** with a regular ventricular rate usually in the high 130's to 140's beats/min [10, 5th ed. pgs. 1082-1083]

- Cardioversion often responds to lower energy levels then Atrial Fibrillation

- Use 50 to 100 J with either a monophasic or biphasic device

[2010 American Heart Association Guidelines for Cardiopulmonary Resuscitation and Emergency Cardiovascular Care, S712]

CARDIOLOGY

ANTIDYSRHYTHMIC DRUGS
Mnemonic: (SOme Block Potassium Channels) [ANK, provided by Dr. Alan Lazzara]

SOme: Sodium Channel Blockers (Class 1) [12, 5th ed., pg 215] [10, 6th ed., pg 1205]	Fast Na+ channel blockers → slow phase 0 depolarization in His-Purkinje and ventricular myocytes = slow conduction and membrane-stabilization Class IA (Quinidine, Procainamide, Disopyramide) Class IB (Lidocaine, Tocainide, Mexiletine, Phenytoin, Aprindine) Class IC (Flecanide, Encainide, Propafenone, Lorcainide) • Toxicity of Class IA, IC meds → ↑ QT → torsade de pointes • Toxicity of Class IB Lidocaine = CNS disturbances (lightheadedness, confusion) CV effects (↓ BP, AV block)
BLOCK: Beta Blockers (Class 2)	Metoprolol, Atenolol, Esmolol, Propanolol, Timolol
POTASSIUM K+ Channel Blockers (Class 3)	Prolong action potential duration and refractory period duration, such that another AP cannot take place immediately following its predecessor; antifibrillatory properties Amiodarone, Dofetilide, Dronedarone, Azimilide Sotalol (non-selective beta-blocker) and Ibutilide also share activity with class II agents
CHANNELS Ca2+ (slow) Channel Blockers (Class 4)	Verapamil, Diltiazem

CAUSES OF PAROXYSMAL SUPRAVENTRICULAR TACHYCARDIA
Mnemonic: (MI PhD) 2 (CREW)

Rate = 130-220, usually 160; Regular

MI	**MI** **M**V Disease (MV Prolapse, MV stenosis)
P	**P**neumonia **P**ericarditis
H	**H**ypertension **H**yperthyroidism
D	**D**igitalis toxicity **D**rop volume (hypovolemia)

C	**C**OPD (much more common rhythm for COPD = MAT)
R	**R**heumatic Heart Disease
E	**E**TOH
W	**W**PW (40-80% PSVT, 10-20% A fib, 5% Flutter) [12, pg. 159]

CARDIOLOGY

PSVT TREATMENT OPTIONS
[2010 American Heart Association Guidelines for Cardiopulmonary Resuscitation and Emergency Cardiovascular Care]

If unstable → "synchronized" Cardioversion → 50 to 100J monophasic or biphasic [AHA: 2011]

Vagal maneuvers → Adenosine (Adenocard) 6 mg IVP → Adenosine 12 mg IVP in proximal vein (short half life); may repeat 12 mg IVP dose once → rhythm does not convert → Narrow or Wide QRS ?

→ Narrow QRS consider → Diltiazem or β-blockers
→ Wide QRS consider → Procainamide or Amiodarone

Methylxanthines (Theophylline) and caffeine → antagonize Adenosine
Dipyridamole (Persantine) and Carbamazepine (Tegretol) → potentiate Adenosine

- Adenosine ultra short acting, 20 secs → AV block; converts >90% of reentrant SVT [12, pg. 148]

- Most common pediatric dysrhythmia = PSVT heart rate usually > 220 (in adults it's less)
  Treatment = Adenosine 0.1 mg/kg (max 6 mg) → 0.2 mg/kg (max 12 mg), (may repeat x1)

- Verapamil is contraindicated in infants

CAUSES OF PULSELESS ELECTRICAL ACTIVITY

Mnemonic: (MI OD PATCH$_6$) [ANK]

MI	MI
OD	**O**verdose

P	**P**E
A	**A**cidosis
T	**T**ension Ptx
C	**C**ardiac tamponade
H6	**H**ypo-xia **H**ypo-thermia **H**ypo-glycemia **H**ypo-volemia **H**ypo-kalemia **H**yper-kalemia

CARDIOLOGY

TREATMENT OPTIONS: ASYSTOLE / PEA
[2010 American Heart Association Guidelines for Cardiopulmonary Resuscitation and Emergency Cardiovascular Care, 8.2]

- Epinephrine 1 mg IVP repeat every 3-5 minutes
- Vasopressin was removed from cardiac arrest algorithm in 2015
- Atropine was removed from the cardiac arrest algorithm in 2010
- Available evidence suggests that the routine use of atropine during PEA or asystole is unlikely to have a therapeutic benefit. (Class IIb, LOE B)[https://eccguidelines.heart.org/index.php/circulation/cpr-ecc-guidelines-2/part-7-adult-advanced-cardiovascular-life-support/]

CAUSES OF PROLONGED QT ON EKG
[23, pg. 130]

- Hypo-calcemia
- Hypo-magnesemia
- Hypo-kalemia
- Hypo-thyroidism
- Hypo-thermia (also, see mnemonics: causes of ST segment elevation and Afib)
- Hereditary
- Drugs
    - TCA's
    - Lithium
    - Phenothiazine
    - Diphenhydramine
    - Cocaine
    - Class IA, IC, III antiarrhythmics
    - Erythromycin IV
    - Zofran (ondansetron)
- Miscellaneous
    - AMI
    - ↑ICP
    - SAH (also, diffuse deep T-wave inversion with SAH)

CAUSES OF LOW VOLTAGE EKG
Mnemonic: (ME BIRP)

M	Myxedema
E	Effusion, pericardial

B	Barrel chest/obese
I	Improper lead placement
R	Restrictive cardiomyopathy
P	Pericarditis / Myocarditis

CARDIOLOGY

FOUR PHASES OF PERICARDITIS
[10, 5th ed, pg. 1132]

Phase 1	ST-Segment ↑ I, V5, V6 → subepicardial ventricular injury
	ST segment in concave upward, in AMI = convex upward
	PR Depression II, aVF and V4 to V6 → subepicardial atrial injury
	Low voltage
	ST-segment ↓ aVR or V1
Phase 2	ST-Segment returns to isoelectric line, and T-wave amplitude decreases (flattens)
Phase 3	T-wave ↓ inversion in leads which were previously ST ↑
	Begins at the end of the second or third week and lasts several weeks
Phase 4	Resolution of repolarization abnormalities

CAUSES OF PERICARDITIS

MNEMONIC: (AMP CARDIAC RIND) [Provided by Dr Mark Postel, with modifications]

A	**A**utoimmune
M	**M**yxedema
P	**P**ost-traumatic (4 to 12 days post-injury)

C	**C**ollagen Vascular Disease (SLE, RA, scleroderma, dermatomyositis, sarcoid, amyloid)
A	**A**ortic dissection
R	**R**adiation = most spontaneously resolve
D	**D**rug-Induced (procainamide, hydralazine, cromolyn sodium, dantrolene, methysergide)
I	**I**diopathic = Most Common; viral = #2 most common cause of pericarditis
A	**A**cute Renal Failure
C	**C**ardiac Infarction

R	**R**heumatic Fever
I	**I**nfectious = viral (coxsackie B, echovirus, influenza, adenovirus, HIV, EBV, CMV), Staph, Strep pneumo, Strep pyogenes (acute rheumatic fever), Mycoplasma, C. trachomatis, Rickettsia, parasites, TB, Salmonella, Haemophilus influenzae
N	**N**eoplasm, mainly metastatic
D	**D**ressler syndrome = several weeks after MI

CARDIOLOGY

PERICARDITIS PEARLS
- Uremia = few EKG changes; → serous or hemorrhagic effusions; hemorrhagic effusions more common secondary to uremia-induced platelet dysfunction
    - Treatment = hemodialysis, steroids 1-2 weeks; avoid NSAIDs → bleeding

- Pericarditis is the most common cardiac manifestation of SLE in 30%
    - Most common symptom = CP which ↑ when the patient is supine, ↓leaning forward
    - Most common physical finding = pericardial friction rub [12, 6th ed., pg 382-383]

CHF CAUSES / PRECIPITATIONS FACTORS / PEARLS
Mnemonic: (HEART MISHAPS) 2 [5, with modifications]

Most common cause of Acute CHF = CAD [10, 5th ed., pg 1053,1115-1118]
Most common causes of Chronic CHF = see below

H	**H**TN	**H**igh output failure vs low output failure*
E	**E**ndocarditis	**E**TOH
A	**A**cute Anemia	**A**rrhythmias
R	**R**heumatic fever	**R**enal failure
T	**T**hyroid (hyper)	**T**ox (Causes of NON-cardiogenic pulmonary edema → see Toxicology)

M	**M**R/MS/AR/AS	**M**yopathies (Dilated / Hypertrophic / Restrictive)**
I	**I**schemia	**I**nfection (Coxsackie B, Diptheria & many more...) Chaga's disease (Trypanosoma cruzi) leading form of CHF in Central America
S	**S**odium load	**S**tiff ventricles (noncompliant)***
H	**H**eat	**H**umidity
A	p**A**gets	**A**nti-inflammatory (NSAIDS)
P	**P**E	**P**regnancy
S	**S**teroids	**S**aline overload

*****High output failure** examples include: hyperthyroid, Paget's, pregnancy, anemia, AV fistula and beriberi (thiamine (B1) deficiency → dilated cardiomyopathy) → heart cannot meet elevated circulatory demands of conditions listed

Low output failure myocardial dysfunction prevents normal metabolic requirements being met

******Cardiomyopathies**
Dilated = infective myocarditis, collagen vascular disease, EOTH, idiopathic, DM, ↑ or ↓thyroidism, uremia and beriberi

Hypertrophic obstructive cardiomyopathy (HOCM) has also historically been known as idiopathic hypertrophic subaortic stenosis (IHSS)

CARDIOLOGY

Restrictive = idiopathic (most common), amyloid, sarcoid and hemochromatosis

***Stiff ventricles /noncompliant - due to hypertrophy or ischemia, requires ↑LVEDP to achieve diastolic filling

CHF ON CXR

Kerley A lines: straight, non-branching lines in the mid and upper lung fields radiate towards the hila, 3 to 4cm long [8]

Kerley B lines: horizontal, non-branching lines seen laterally in lower zones never > 2 cm long

Pulmonary edema, pleural effusions (usually bilateral with right > left, if unilateral almost always right sided), cardiac enlargement without specific chamber enlargement and evidence of ↑pulmonary venous pressure (enlargement of vessel in upper lung zones "**cephalization**") [8]

TREATMENT OPTIONS IN CHF
Mnemonic: (L M N O P)

L	**L**asix = 2 mechanism of action: 1) Vasodilation within few minutes 2) Diuretic response in 10-20" • Dose = equal twice the patient's daily usage, up to a maximum of 180 mg IV • If the patient previously has not been on a loop diuretic, initial dose = 40mg IV • If bumetanide (Bumex) is used, 1 mg of Bumetanide equals 40 mg furosemide • Ethacrynic acid is useful if the patient has a serious **sulfa** allergy
M	**M**orphine 2-6 mg IV
N	**N**itroglycerin 0.4 mg SL q 1-5 min, IV starting dose = 0.2-0.4mcg/kg/min
O	**O**xygen, CPAP, intubate
P	**P**osition – elevate HOB

- CHF occurs if there is acute impairment of at least 25% of the left ventricle

- Most common symptom of left-sided failure = SOB

- The most common and earliest finding on CXR in CHF = cephalization
  CXR findings may precede symptoms

- LEFT sided failure (reduced flow into aorta and systemic circulation) → fatigue, dyspnea, orthopnea, ↑HR, ↑RR, ↑BP, ↑CO, S3 gallop, diaphoresis, rales/wheezing

- RIGHT sided failure (secondary to elevated systemic venous pressure) → JVD, ↓BP, ↓CO, RUQ pain, peripheral edema, hepatomegaly and hepatojugular reflux

CARDIOLOGY

ABDOMINAL AORTIC ANEURYSM (AAA)

- AAA = localized dilation of aorta, involves all three layers of aorta (intima, media and adventitia ... don't confuse with Aortic Dissection

- Aortic Dissection = blood enters the media of the aorta and dissects the aortic wall

- Men > Female

- Average age at time of diagnosis = 65 to 70 y/o (75% of AAAs occur in patients > 60 y/o)

- AAAs have traditionally been attributed to Atherosclerosis but other factors probably contribute to their formation → biochemical abnormalities and uncertain genetic basis → Society for Vascular Surgery recommend labeling AAA as "nonspecific" rather than "atherosclerotic" [10, 6th ed., pg. 1331]

- Strong risk factor for AAA = Family history → patients with 1st degree relative & AAA have 10-20- fold ↑risk of developing AAA [10, 6th ed., pg. 1331]; other risk factors = smoking, HTN, history of CAD or PVD

- Most important risk factor for AAA rupture = size; most ruptures > 5cm [10, 6th ed., pg. 1331]

- KUB /cross-table lateral = suggest diagnosis of AAA in > 65% of symptomatic patients (Calcified aorta/loss of renal/psoas shadow) [12, 5th ed., pg. 414]

- Classic triad for ruptured AAA = ↓BP, pulsatile abdominal mass, and flank/back pain
  Triad may be incomplete in as many as 50%

- Mechanism of pain = 1. Rapid expansion of AAA or 2. Pressure of AAA on surrounding structures like nerves or 3. Presence of free blood in the abdomen or retroperitoneum from leaking or ruptured aneurysm

- Most common site AAA ruptures = Retroperitoneal 75-90% of cases (most common on left)
  Most common segment = below renal arteries
  10 to 30% rupture intraperitoneal = if intraperitoneal → rapidly fatal [10, 6th ed., pg. 1331]

AORTIC DISSECTION

- Thoracic aortic dissection is the most common lethal disease affecting the aorta
- 2-3x more common as ruptured AAA [10, 4th ed., pg 1819] [EM Reports, Vol. 21, #1]
- Major risk factor for aortic dissection = HTN [10, 5th ed., pg 1324]
  Other = pregnancy, aortic valve disease, connective tissue disease, stimulant use
- Age > 60 y/o; 3:1 male:female
- Type A (proximal) = ascending aorta +/- descending aorta = 75% cases = surgery
  Type B (distal) = confined to descending aorta = medical management
- BP difference between upper extremities of > 20 mmHg or a loss/reduction of lower extremity pulses suggests aortic dissection

CXR Findings
- Widened mediastinum (> 8cm) - most common/reliable abnormality occurs 75% of cases
- Tracheal deviation to the RIGHT
- Depressed LEFT mainstem bronchus
- LEFT apical cap
- Loss of aortic arch

CARDIOLOGY

- Separation (> 5mm) of intimal calcium from the outer border of the aorta
- LEFT pleural effusion (see mnemonic "causes of pleural effusions")

Cardiovascular [10, 5th ed., pg 1325]
- 90% CP; Migration of pain down back or from chest to back is highly specific for dissection
- Ascending dissections → pain anterior chest
- Arch dissections → neck and jaw pain
- Descending dissections → pain in the interscapular area
- Aortic Insufficiency (diastolic murmur) = common > 50% of patients → Acute CHF
- EKG is abnormal in most patients – varying degrees of heart block, AMI
- Tamponade (blood spreads proximally to open the pericardial space to aortic blood flow)

Neurologic
- 9% syncope (most common cause = tamponade; other = cerebral ischemia) [10, 5th ed., pg 1325]
- Type A→ carotid artery dissection/obstruction → CVA
- Spinal artery of Adamkiewicz blockage → cord ischemia → acute paraplegia
- Horners syndrome = compression of sympathetic chain from enlarged aorta

GI
- Hematemesis
- Accompanied mesenteric ischemia BAD → death rate 88%

Renal
- Hematuria, oliguria
- Decrease blood flow → renin release → refractory ↑BP

HYPERTENSIVE EMERGENCY

Severe elevation in BP accompanied by acute target organ damage; treatment = parenteral therapy

$$MAP = CO \times SVR \qquad MAP = DBP + 1/3 (SBP - DBP) \qquad CPP = MAP - ICP$$

HYPERTENSIVE EMERGENCIES – TREATMENT OPTIONS

1. HTN encephalopathy = Nitroprusside or Labetalol or Clevidipine (Cleviprex, new CCB drug, recently approved, may see but will be institution dependent)

2. Stroke syndromes = Labetalol or Nicardapine (Cardene)
   Nimodipine (Nimotop) useful in SAH, 60mg po (Note: Patients with acute ischemic stroke-in-evolution are most often not given antihypertensive drugs unless they are candidates for tissue plasminogen activator and their initial blood pressure is ≥185/110 mmHg [http://www.uptodate.com/contents/evaluation-and-treatment-of-hypertensive-emergencies-in-adults]

3. Acute coronary insufficiency = NTG
   Avoid Nitroprusside because of "coronary steal" syndrome [12, 5th ed, pg. 407]

4. Acute pulmonary edema treatment = NTG and Lasix; Drugs that increase cardiac work (eg, hydralazine) or acutely decrease cardiac contractility (eg, labetalol or other beta blockers) should be avoided [http://www.uptodate.com/contents/evaluation-and-treatment-of-hypertensive-emergencies-in-adults]

CARDIOLOGY

5. Aortic Dissection = Nitroprusside + esmolol or labetalol, give β-blocker first to prevent reflex tachycardia. Goal - decrease aortic pressure wave contour (dp/dt); ↓ HR 60

6. Eclampsia = Hydralazine; MgSO4 for seizures. Avoid Nipride → fetal cyanide toxicity

7. Hyperadrenergic states (MAOI, Pheochromocytoma, Anit-HTN withdrawal)
   Phentolamine + Propranolol; Nitroprusside + Propranolol; Labetalol

Trimethaphan (ganglion blocker) if Beta Blocker contraindication

Do not drop BP too quickly. Ischemic damage can occur in vascular beds that have grown accustomed to the higher level of blood pressure (ie, autoregulation). For most hypertensive emergencies, mean arterial pressure should be reduced by about 10 to 20 percent in the first hour and then gradually during the next 23 hours so that the final pressure is reduced by approximately 25 percent compared with baseline.[http://www.uptodate.com/contents/evaluation-and-treatment-of-hypertensive-emergencies-in-adults]

Dosing
- Labetalol = 20mg IV over two minutes – repeat or double every 10min to max of 300mg; Infusion rate 1 to 2 mg/min up to 8 mg/min
- Nicardipine (Cardene) = 5mg/hr, increase infusion 2.5mg/hr every 10 min to a max of 15mg/hr
- Nitroprusside = start at 0.5 mcg/kg/min IV infusion

Hypertensive Urgency
Severe elevations in BP (DBP > 115), with mild or no acute target organ damage; may reduce BP within hours to days usually with oral medications

HTN without Emergency / Urgency = does not mandate urgent therapy
- Essential HTN = chronic problem and referral to PMD will greatly enhance management of HTN
- Treatment of HTN as an outpatient basis is usually not the responsibility of EM doc [10, 5th ed. 1166-1177]

CARIDOLOGY PEARLS

- Most common complications of Anterior Wall MI = Mobitz type II, ventricular aneurysm, CHB (unlike CHB with IWMI, CHB with AWMI = grave prognosis) [12, 5th ed., pg. 363]

- Most common complications of Inferior Wall MI = Acute MR, ↓HR and ↓BP
  First degree AVB, Mobitz type I (Wenckebach, accounts for 90% of 2nd-degree AVB and AMI) and CHB (stable, usually resolves) [12, 5th ed., pg. 363]

- VF occurs in 5% of patients with AMI, 80% present within 12 hours [10, 6th ed., pg. 1160]

- Most common cause of VT/VF = AMI / ischemia

- Low serum potassium, but not magnesium, has been associated with ventricular arrhythmias in AMI
  Maintain serum potassium >4 mEq/L and magnesium >2 mEq/L [2010 AHA Guidelines for Cardio Resusc and Em Cardiovascular Care]

- Most common conduction disturbance in AMI = First degree AV block, 15%; more common with IWMI

CARDIOLOGY

- Mortality rate with IWMI and CHB = 15%; if RV involved MR rises to > 30% [12, 5th ed., pg. 363]

- Most common rhythm disturbances in AMI = PVC's (>90%), PACs (50%) Sinus tach (33%) [12, 5th ed., pg. 362]

- % of patients with AMI who have diagnostic changes on their first EKG = 50% [12, 5th ed., pg. 359]

- Most common complications of thrombolytic therapy in patients with AMI = Reperfusion arrhythmias; accelerated idioventricular rhythm most common arrhythmia

- Most common rhythm disturbance in digitals toxicity = PVC's 60% > SVT 25% > AV block 20%

- Earliest EKG findings associated with AMI = hyperacute T → "giant" R wave ("tombstone") → typical ST segment elevation – which is either flat (horizontally or oblique) or convex

- ST-segment elevation in V1, in the absence of ST-segment elevation in the other anteroseptal leads (V2-V3), is suggestive of right-ventricular ischemia/infarction

- Dressler's syndrome = fever, pleuritis, leukocytosis, pericardial friction rub, and evidence of pericarditis or pleural effusion occurring several weeks after MI. Autoimmune. Treatment = NSAIDs or ASA [10, 6th ed., pg. 1283]

- LV dysfunction > 40% = Cardiogenic shock [10, 6th ed., pg. 47]

- Cardiogenic shock may not occur immediately post-MI. The median delay from AMI to clinical development of cardiogenic shock may be up to 7 hours

- Most common type of CVA after MI = ischemic thromboembolic [10, 6th ed., pg. 1161]

- Most common permanent pacer dysfunction = oversensing

CARDIOLOGY PEARLS - INFECTIVE ENDOCARDITIS

- In patients with rheumatic heart disease, the mitral valve is the most common site of IE involvement

- Most common cause of Native Valve Infective Endocarditis (IE) with valvular or Congenital HD =
    - *Strep viridans* (30-40%)
    - *Staph aureus* + *S. epidermidis* = 20-35%,
    - "Other" Strep (15-25%)
    - *Enterococci* (5-18%)
    - Treat = Pen G or AMP + Nafcillin + Gentamycin [Sanford, 2009, pg. 25]. Most textbooks = AMP + Gent
    - Gentamycin for synergy; Nafcillin does not cover Enterococci
    - or Vancomycin + Gentamycin [Sanford, 2009, pg. 25]

- Most common cause of Native Valve Right-Sided IE = *Staph aureus* (IVD abusers)
  Most common valve affected = tricuspid; Treatment = Vancomycin [Sanford, 2009, pg. 25]

- Culture negative IE = *Haemophilus parainfluenzae, H. aphrophilus, Actinobacillus, Cardiobacterium, Eikenella,* and *Kingella* species (HACEK organisms); Treatment = Ceftriaxone 2gm IV q 24hrs or Amp + Gent

CARDIOLOGY

Add *Bartonella henselae, B. quinata* = (HABCEK); Treatment = Ceftriaxone + Gentamycin + Doxy
Also, *Coxiella burnetii* Treatment = Doxy + Hydroxychloroquine [Sanford, 2009, pg. 27] and [10, 6th ed., pg. 1301]

- Most common cause of Prosthetic Valve IE < 60 days post-op = coagulase negative *Staph. (S. epidermidis)* and *Staph aureus*. Treat = Vancomycin + Gentamycin + Rifampin [Sanford, 2009 pg. 28]

- Most common cause of Prosthetic Valve IE > 60 days post-op = similar to Native Valve with Staph aureus now most common etiology [Sanford, 2009 pg. 28, JAMA 297:1354,2007]

- Association between *Streptococcus bovis* IE and coexisting GI malignancy [10, 6th ed., pg. 1301]

- IE = petechiae, splinter hemorrhages (dark red vertical lesions in nailbeds), Osler nodes (painful, red, raised lesions on distal finger pads), Janeway lesions 35% (flat, red-bluish, painless lesions on palms/soles), fever, murmur, anemia, malaise, Roth spots (retinal hemorrhages with pale center)
Fever is the most common presenting symptom

- Most common cause of Acute Aortic Regurgitation (AR) = Infective endocarditis

- In developed countries, the frequency of RHD has declined, and MVP is now the most common underlying condition in patients with endocarditis

INFECTIVE ENDOCARDITIS
Mnemonic (FROM JANE)

F	**F**ever (most common presenting symptom)
R	**R**oth spots (retinal hemorrhages with pale center)
O	**O**sler nodes (painful, red, raised lesions on distal finger pads)
M	**M**urmur

J	**J**aneway lesions 35% (flat, red-bluish, painless lesions on palms/soles) and petechiae
A	**A**nemia
N	**N**ailbeds - splinter hemorrhages (dark red vertical lesions in nailbeds)
E	**E**mboli

CARDIOLOGY

CARDIOLOGY PEARLS - VALVULAR HEART DISEASE

- Most common cause of Chronic Aortic Regurgitation (AR) = Rheumatic heart disease (RHD)

- Most common cause of Aortic Stenosis < 70 y/o=Congenital Heart Disease (Bicuspid Valve 50%), 2nd RHD

- Most common cause of Aortic Stenosis > 70 y/o=Idiopathic calcification/degen. heart disease [10, 5th ed. pg. 1155]

- Most common cause of Acute MR = Rupture of chordae tendineae or papillary muscle (IWMI)

- Most common cause of Chronic MR = Rheumatic heart disease (RHD)

- Most common cause of Mitral Stenosis (MS) = Rheumatic heart disease (RHD)

- Most common presenting symptom of all cardiac valvular diseases = exertional dyspnea

- Most common symptom of Mitral Stenosis (MS) = Exertional Dyspnea (symptom specific to MS = hemoptysis)

- Aortic Stenosis (AS) = exertional Syncope, Angina and Dyspnea (mnemonic=SAD)

- Most common rapidly lethal complication of Aortic Stenosis (AS) = sudden death

- Most common valvulopathy due to chest trauma = Aortic Regurgitation (AR)

- Aortic Regurgitation (AR) = High Pulse Pressure, head bobbing = prominent ventricular impulse (Musset sign), soft diastolic murmur (Austin-Flint murmur), bounding peripheral pulses (water hammer), pulsations of uvula and nailbed, and SBP of LE > UE (Hill sign); causes: trauma, IE, aortic dissection, Marfan's, syphilis

- Systolic murmurs = AS (mid-systolic ejection crescendo-decrescendo) and MR (holosystolic)

- Diastolic murmurs = AR (blowing decrescendo) and MS (holodiastolic with opening snap) [http://www.wilkes.med.ucla.edu/Systolic.htm]

- MS, MR, Pulmonary Insufficiency = decrease pulse pressure (AR high pulse pressure) [10, 5th ed., pg 1155]

- Most common symptoms of MVP = CP (sharp, localized) & palpitations; affects 10% of population; majority asymptomatic; EKG abnormal; patients may have pectus excavatum or scoliosis; ↑ migraines, anxiety and CVA

  ↑incidence of sudden death and dysrhythmias. ↑incidence of TIAs under the age of 45; MVP with regurgitation (both leafltets involved; usually affects one – the posterior leaflet) → ↑risk of IE [10, 5th ed., pg. 1153]

- MVP murmur = mid-systolic click (snapping of chordae tendineae during prolapse of valve) followed by late systolic crescendo murmur heard best at apex or LSB with patient in left lateral decub; If ↓ preload (↓EDV) (valsalva, standing position) → click moves closer to S1 → ↑ previously unheard click, murmur is longer not louder; handgrip → ↑ louder murmur [10, 5th ed., pg. 1153]

CARDIOLOGY

- Valsalva → ↑intrathoracic pressure → ↓venous return → ↓preload → ↓most murmurs except for Hypertrophic Cardiomyopathy murmur which ↑ because dynamic LV outflow obstruction is accentuated by ↓preload

- Hypertrophic Cardiomyopathy murmur = harsh mid-SEM crescendo-decrescendo; loud S4; heard best at apex & left sternal border; does not radiate to neck (AS radiates to carotids) [10, 5th ed., pg. 1144]
  ↓ preload (standing, valsalva diuretics or nitrates) or ↓ afterload (vasodilators) →↑gradient → ↑murmur

CARDIOLOGY PEARLS

- Amount of fluid in normal pericardial space = 25-50 ml

- Need 250 ml of fluid in pericardial space before cardiac silhouette ↑ on CXR

- Beck's Triad: 1) Muffled heart tones 2) ↓BP 3) JVD (↑CVP) = cardiac tamponade

- Most common echocardiographic findings with cardiac tamponade = right ventricular diastolic collapse

- Electrical alternans on EKG is pathognomonic for tamponade

- Most common cause of pericardial tamponade = malignancy 30-60%; uremia 10-15%, idiopathic pericarditis 5-15%, ID 5-10%, anticoagulation 5-10%, connective tissue diseases 2-6%, and Dressler syndrome 1-2%

- Causes of pulsus paradoxus = cardiac tamponade, and obstructive lung disease (asthma, COPD)

- Pulsus Paradoxus = measuring the variation of SBP during expiration and inspiration [10, 5th ed., pg 398]

  Slowly decrease cuff pressure until systolic sounds are first heard during expiration but not during inspiration, (note this reading)
  Slowly continue decreasing the cuff pressure until sounds are heard throughout the respiratory cycle, (inspiration and expiration) = note this second reading

  If the pressure difference between the two readings is >10mmHg = pulsus paradoxus

- Most common cause of restrictive cardiomyopathy = idiopathic [10, 5th ed. pg. 1145]
  Other causes = amyloid, sarcoid and hemochromatosis

- Most common infectious cause of myocarditis = Coxsackie B
  Consider CMV & Toxoplasma gondii in Transplant or HIV patient
  Kawasaki = 50% myocarditis
  Chaga's disease (Trypanosoma cruzi) = leading cause of death in Central America; 3/4 no symptoms
  Spread by the reduviid, "kissing" or "assassin" bug
  Pathognomonic finding in Chaga's = Romaña sign = painless unilateral periorbital edema; uncommon
  Chagoma = painful cutaneous edema at the site of skin penetration
  Treatment = Nifurtimox (Lampit) or Primaquine

- Shock = imbalance of tissue O2 supply and demand;
  All patients with shock should receive as the first priority → Supplemental oxygen

CARDIOLOGY

- Four mechanistic classifications of shock [12, 5th ed., pg 215]
  1) Hypovolemic (inadequate circulatory volume)
  2) Cardiogenic (inadequate cardiac pump function)
  3) Distributive (peripheral vasodilation and maldistribution of blood flow); examples → neurogenic shock (↓BP and ↓HR), anaphylactic shock, pancreatitis, burns, trauma, adrenal insufficiency, drug or toxin reactions, heavy metal poisoning, hepatic insufficiency
  4) Obstructive (extra-cardiac obstruction to blood flow) cardiac tamponade, PE, tension pneumothorax

- Pure alpha adrenergic agent = phenylephrine

- Mixed alpha and beta adrenergic agents = epinephrine, norepinephrine and dopamine

- Pure beta or primary beta-agonists = dobutamine and isoproterenol

- S1Q3 RAD = Left Posterior Fascicular (LPFB)
  S3Q1 LAD = Left Anterior Fascicular (LAFB)

- LPFB has more serious implications since it implies compromise to both the right and left coronary arteries as well as damage to large areas of myocardial muscle and to the electrical conduction system in the left ventricle [Sensible analysis of the 12-lead ECG By Kathryn Monica Lewis, Kathleen A. Handal, pg. 170]

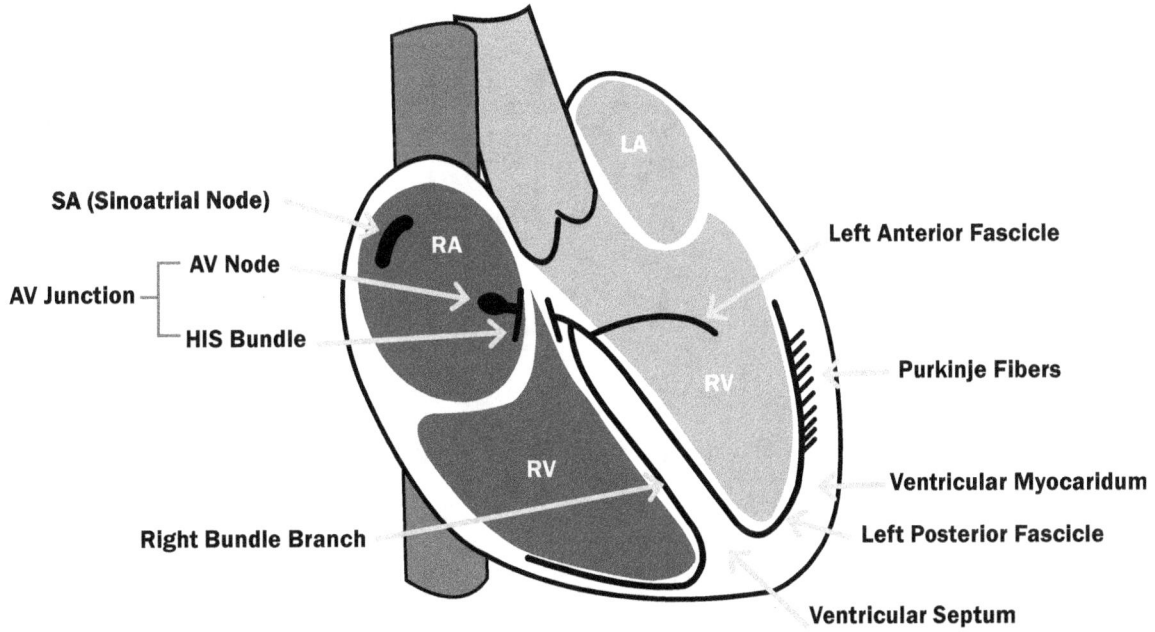

CARDIOLOGY

PULMONARY

RISK FACTORS FOR DVT / PULMONARY EMBOLISM
Mnemonic: (MOIST CAMEL) [3, w/modification]

M	**M**obility = Stasis (prolonged travel, bed rest, paralysis (CVA, spinal cord injury), leg cast
O	**O**besity
I	**I**nflammatory Conditions (IBD, SLE, PVD)
S	**S**urgery (especially orthopedic, pelvic & major abdominal surgery)
T	**T**rauma = Intimal damage (trauma, IVDA, surgery, central lines)

C	**C**HF / CVA / COPD (respiratory diseases) / Cardiac Disease
A	**A**ntithrombin III, Protein C & S, deficiency
M	**M**alignancy / Medical patients = (MEDENOX) trial
E	**E**strogen / pregnancy /postpartum /elective AB or miscarriage
L	**L**ong bone fracture

VIRCHOW'S TRIAD
1) Hypercoagulable state
2) Venous stasis
3) Venous injury

HYPERCOAGULABLE (THROMBOPHILIA) STATES
- Previous DVT/PE
- Nephrotic syndrome (loss of antithrombin)
- Malignancy
- Inflammatory Conditions (IBD, SLE, PVD)
- Sepsis

COAGULATION DISORDERS – INHERITABLE VS ACQUIRED
- Protein C or S deficiency
- Resistance to activated Protein C
- Antithrombin deficiency
- Disorders of Fibrinogen or Plasminogen
- Antiphospholipid antibodies (lupus anticoagulant & anti-cardiolipin)

INCREASED ESTROGEN (CAUSES URINARY LOSS OF PROTEIN S AND ANTITHROMBIN)
- Pregnancy
- Postpartum status < 3 months
- OCPs
- Elective abortion or miscarriage

PULMONARY

D-DIMER

Fibrin fragments found in fresh fibrin clot & in fibrin degradation products. Elevated in many conditions (poor Sensitivity 77%) including = DVT/PE, CVA, trauma, cancer, surgery, infection/sepsis, sickle cell anemia, postpartum within 1 week and pregnancy (elevated in 75% of patients with a normal pregnancy)

WELLS CLINICAL SCORE FOR DVT

Clinical Parameter Score	Score
Active cancer (treatment ongoing, or withing 6 mo. or palliative	+1
Paralysis or recent plaster immobilization of the lower extremities	+1
Recently bedridden for >3 days or major surgery < 4wks.	+1
Localized tenderness along the distribution of the deep venous system	+1
Entire leg swelling	+1
Calf swelling >3 cm compared with the asymptomatic leg	+1
Previous DVT documented	+1
Collateral superficial veins (non varicose)	+1
Alternative diagnosis (as likely or greater than that of DVT)	-2
Total of Above Score	
High probability	≥ 3
Moderate probability	1 or 2
Low probability	0

D-DIMER + WELLS CLINICAL SCORE FOR DVT- DIAGNOSTIC STRATEGY

A negative D-dimer result in the *unlikely* group (Wells DVT score < 2) ----> rules out DVT
All patients with a positive D-dimer result and all patients in the *likely* group (Wells DVT score > 2) ----> require a diagnostic study (duplex Ultrasonography)

ANTICOAGULATION TREATMENT OPTIONS FOR VENOTHROMBOEMBOLISM (VTE) - DVT/PE

1. Injectable indirect factor Xa and IIa (thrombin) inhibitors
   The effects of indirect inhibitors are mediated through antithrombin (AT)
   a. Unfractionated heparin (UFH); Xa and to a lesser extent IIa (thrombin) Also, inhibits factors XIIa, XIa and IXa
   b. Enoxaparin (Lovenox), low-molecular-weight-heparin (LMWH); Xa and to a lesser extent IIa (thrombin) inhibitor
   c. Fondaparinux (Arixtra); exclusive indirect factor Xa inhibitor

2. Target-Specific Oral Anticoagulants (TSOACs) or Direct Oral Anticoagulants (DOACs)
   a. Direct Thrombin Inhibitors (DTIs)
      i. Dabigatran (Pradaxa)
   b. Direct Factor Xa Inhibitors ("Xabans")
      i. Rivaroxaban (Xarelto)
      ii. Apixaban (Eliquis)
      iii. Edoxaban (Lixiana, Savaysa)

PULMONARY

VTE PROPHYLAXIS

UFH 5,000 units SQ every 8 hours

Enoxaparin (Lovenox)
- 40 mg SQ once daily or 30 mg SQ once daily if CrCl < 30 mL/min

Fondaparinux (Arixtra)
- 2.5 mg SQ daily if CrCl > 30 mL/min and actual body weight > 50kg

Rivaroxaban (Xarelto) - Surgical VTE Prophylaxis
- Knee replacement - 10 mg once daily for 12 days
- Hip replacement - 10 mg once daily for 35 days
- Initial dose should be taken at least 6-10 hours after surgery once hemostasis has been established
- Avoid use if CrCl < 30 mL/min

Apixaban (Eliquis) - Surgical VTE Prophylaxis
- Knee replacement – 2.5mg BID for 12 days
- Hip replacement - 2.5mg BID for 35 days
- Initial dose should be given 12 to 24 hours after surgery once hemostasis has been established
- No dose adjustment with moderate renal impairment for above Apixaban VTE indications
- (not studied in patients with CrCl < 25 mL/min)

VTE TREATMENT

UFH Bolus 80 units/kg followed by infusion of 18 units/kg/hr

Enoxaparin (Lovenox)
- 1 mg/kg SC every 12 hours or
- 1.5 mg/kg SC daily - "suggested over twice-daily regimen"

Fondaparinux (Arixtra)
- Weight <50 kg: 5 mg SQ daily
- Weight 50-100 kg: 7.5 mg SQ daily
- Weight >100 kg: 10 mg SQ daily

ARIXTRA is contraindicated:
- Creatinine clearance <30 mL/min
- DVT prophylaxis if weight < 50 kg

Rivaroxaban (Xarelto)
- 15 mg BID x 21 days
- On day #22 transition to
- 20 mg once daily

Reduce Risk of Recurrent DVT or PE
- Following 6 months of treatment
- 20 mg once daily

PULMONARY

Apixaban (Eliquis)
- 10 mg BID x 7 days, then 5 mg BID

Reduce Risk of Recurrent DVT or PE
- Following 6 months of treatment
- 2.5 mg BID

Edoxaban (Lixiana, Savaysa)
- >60 kg: 60 mg daily
- <60 kg: 30 mg daily
- Treat with parenteral anticoagulation for 5-10 days (dual therapy)
- CrCl 15 to 50 mL/min: decrease dose to 30 mg daily

Dabigatran (Pradaxa)
- 150 mg BID
- Treat with parenteral anticoagulation for 5-10 days (dual therapy)
- Avoid use CrCl < 30 mL/min

TREATMENT STRATEGIES FOR MAJOR BLEEDING FROM TSOACS

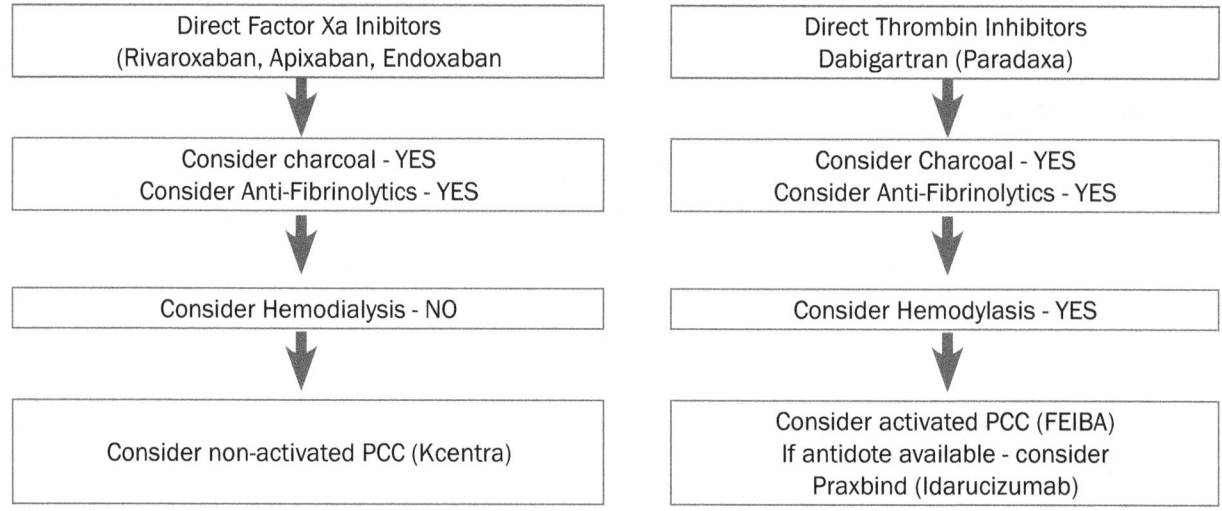

Praxbind (Idarucizumab)
Idarucizumab is a humanized, monoclonal, antibody fragment that specifically binds with high affinity to dabigatran. Dabigatran has an affinity for idarucizumab that is 350 times greater than its affinity for thrombin.

Andexanet Alfa
Antidote in pipeline for oral direct and injectable indirect Factor Xa InhibitorsAndexanet alfa is a modified recombinant factor Xa molecule administered intravenously. Antidote to reverse the anticoagulant activity of oral direct (apixaban, edoxaban, and rivaroxaban) and injectable indirect (enoxaparin and fondaparinux) factor Xa inhibitors. Andexanet alfa acts as a decoy to target and sequester with high specificity both oral and injectable factor Xa inhibitors.

PULMONARY

ANTICOAGULATION CASCADE
Contact Activation (intrinsic) Pathway

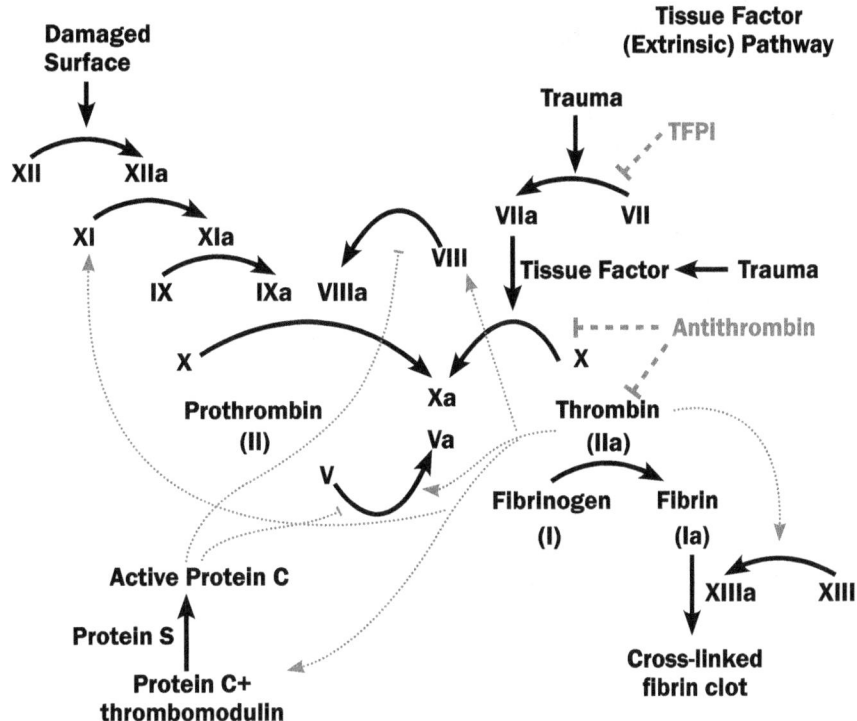

MECHANISM OF ANTICOAGULATION

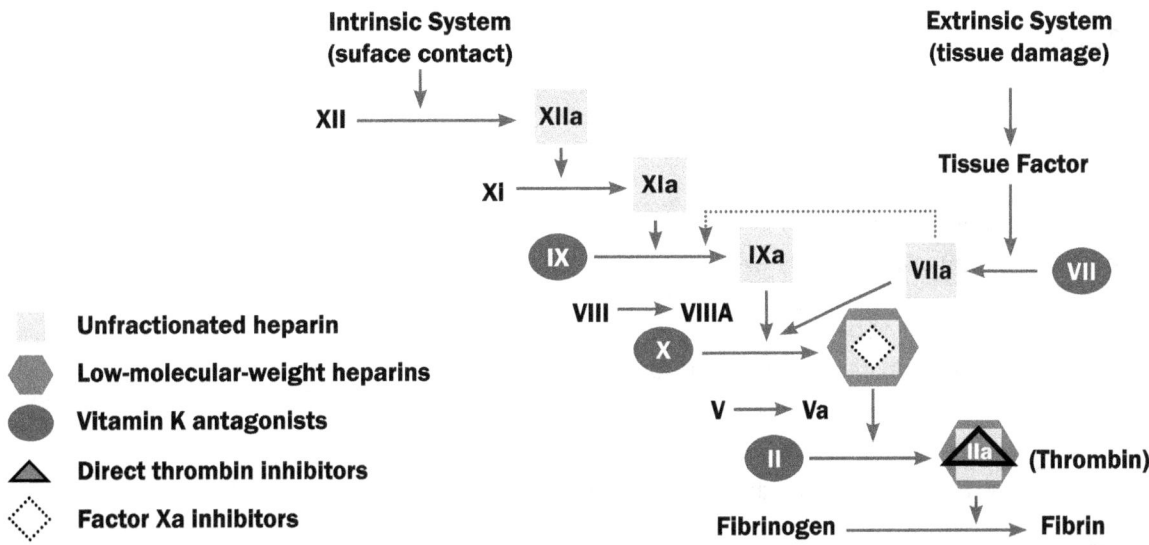

PULMONARY

WELL'S CRITERIA FOR ASSESSMENT OF PRETEST PROBABILITY FOR PE

Clinical Signs and Symptoms of DVT	3 points
An alternative diagnosis in less likely than PE	3 points
Heart Rate > 100	1.5 points
Immobilization at least 3 days, or Surgery in the Previous 4 weeks	1.5 points
Previous, objectively diagnosed PE or DVT	1.5 points
Hemoptysis	1 point
Malignancy w/ Treatment within 6 months, or palliative	1 point

Low-risk patients, score < 2 had a mean probability of **3.6%** for PE
Intermediate-risk patients, score 2-6 had a mean probability of **20.5%** for PE
High-risk patients, score > 6 had a mean probability of **66.7%** for PE

Modified Wells Criteria PE
If < 4 Proceed to PERC
If > 4 CT angio or V/Q

PULMONARY EMBOLISM RULE-OUT CRITERIA (PERC)

- Age >= 50y
- Tachycardia, HR > 100
- O2 sat < 95%
- Prior DVT or PE
- Recent trauma or surgery
- Hemoptysis
- Exogenous estrogen (hormone use)
- Unilateral leg swelling

If NO criteria are met, PE can be ruled out
If any criteria are met, unable to rule out PE

PERC RULE
MNEMONIC - (HAD CLOTS)

The patient must meet the following:

H	no **H**ormones
A	no **A**ge > 50
D	no **D**VT/PE history

C	no **C**oughing blood
L	no **L**ower extremity swelling unilaterally
O	no **O**2 saturation <95
T	no **T**achycardia >100
S	no **S**urgery/trauma within past 28 days

PULMONARY

PULMONARY EMBOLISM - PEARLS

- Most Common **Symptom** = dyspnea, 90% > CP [10, 6th ed., pg 1371-1372]

- Most Common **Sign** = tachypnea, 70%

- Triad of Dyspnea, CP and Hemoptysis < 20%

- Most common cause of death from pulmonary thromboembolism = hemodynamic collapse [10]

- Most common **EKG** finding = Non specific ST / T Wave Changes
  (50% of patients diagnosed with PE have a HR **< 100** beats/min [12, 6th ed., pg 388]

EKG FINDINGS IN PULMONARY EMBOLISM
Mnemonic – (A STRIPS)

| A | Atrial fibrillation, new onset (see mnemonic AFIB) |

S	S1Q3T3
T	Tachycardia
R	RAD, RBBB
I	Inverted T-wave in V1-V4
P	Pulmonale (peaked P waves in lead II)
S	ST segment elevation

The CXR is most often abnormal in PE

- Conflicting literature:
  Most common CXR finding from PIOPED trail = atelectasis
  Most common CXR finding from UPET (Urokinase PE Trail) = elevated hemidiaphragm

- Most common CXR in admitted patients = cardiomegaly [12, 6th ed., 390]

- Most common CXR in outpatient patients = basilar atelectasis [12, 6th ed., 390]
- Other CXR findings: Hampton's hump, (wedge-shaped consolidation in lung periphery), Westermark's sign, (dilation of proximal arteries with collapse of distal vasculature)

PULMONARY EMBOLISM - PEARLS

- **ABG** = acute respiratory alkalosis, hypoxemia, an abnormal AaO2 gradient or normal

- **A-a gradient** = should be widened; Sn 90%, Sp 15%

- A-a gradient = 150 – PaO2 – (PaCO2 x 1.25)
  Normal = 5-20 (calculate on room air)

PULMONARY

- Lack of hypoxemia does NOT rule out diagnosis of PE

- Hypoxia helps with risk stratification → correlates with degree of pulmonary vasculature occlusion from PE → shown to predict poor outcome [12, 6th ed., 389]

- Consider spiral **CT** of chest, **V/Q**, or pulmonary angiography

- ECHO = right heart strain in 40%
  McConnell's sign = RV hypokinesis in the presence of normal RV apical contractility

PULMONARY EMBOLISM – PEARLS PREGNANCY

- Threshold for human teratogenesis = 10 rad; fetus most vulnerable 8 to 15 weeks gestation
  [12, 6th ed., pg. 675]

- V/Q — Total fetal exposure to xenon-133 and technetium-99m = 0.5 rad; void bladder x 3

- CXR = 0.00005 rad; CT head < 0.1 rad, **CT chest = < 1 rad**; CT abdomen = 3.5 rad
  [12, 6th ed., pg. 675]

- British Journal of Radiology [2006] [79, 441-444] recommends:
  D-dimer → positive → bilateral lower extremity venous doppler → non-diagnostic → CT chest over V/Q to diagnose PE

TREATMENT OPTIONS IN PULMONARY EMBOLISM
[10, 6th ed, pgs. 1375-1380] and [Circulation. 2005;112:e28-e32 Management of Massive Pulmonary Embolism]

American College of Chest Physicians Evidence-Based Clinical Practice Guidelines (9th Edition). CHEST 2012;141(2)(Suppl):7S-47S.

- **Heparin**: bolus (80U/kg or 5,000 Units) then infusion (18U/kg/hr or 1,300 u/hr)

- In massive PE consider 10,000 U bolus, followed by a continuous intravenous infusion of at least 1,250 U/h (however 80U/kg bolus followed by 18U/kg/hr, infusion acceptable)

- **LMWH**: Enoxaparin (Lovenox) = 1mg/kg SQ or IV every 12 hours or 1.5mg/kg q 24 hours
  Circulation. 2005;112:e28-e32 Management of Massive Pulmonary Embolism did not mention Lovenox as a treatment option for "Massive" PE

- **tPA** = You will find three different protocols for tPA
  - 100 mg over **2 hours** (FDA approved regimen, most textbooks) *or*
  - 15 mg bolus, then 85 mg continuous infusion over **2 hours** or
  - Accelerated infusion regimen used in AMI

PULMONARY

Hold **heparin** during fibrinolytic infusion. At the conclusion of alteplase infusion begin heparin infusion without a bolus when aPTT has decreased to < 80 seconds.

- Hypotension
    - 0.9% NS cautiously because ↑ RV wall stress can → ↓ the ratio of RV oxygen supply to demand → may result in ischemia, deterioration of RV function, and worsening RV failure and → further ↑interventricular septal shift toward the left ventricle, thereby worsening left ventricular compliance and filling. Consider only 500 to 1,000 cc

    - Norepinephrine and dobutamine → permits increased myocardial contractility, while minimizing both vasodilation and the risk of hypotension

- Morphine 4 to 6mg IV PRN

Massive PE = arterial hypotension + cardiogenic shock
- Hypotension: SBP < 90 mm Hg or drop in SBP of at least 40 mm Hg for at least 15 minutes
- Shock = tissue hypoperfusion & hypoxia, altered LOC, oliguria, or cool, clammy extremities

TREATMENT OPTIONS IF MAJOR BLEEDING STARTS AFTER THROMBOLYTIC THERAPY

- Stop the lytic agent
- FFP and cryoprecipitate ASAP
- Aminocaproic acid (Amicar) Inhibits plasminogen activators - 5 gm bolus infused over 1 hr then infusion of 1 gm/hr until bleeding has stopped; Rapid administration may result in hypotension, bradycardia and arrythmias. [10, 5th ed., pg. 1229]

TREATMENT OPTIONS IF MAJOR BLEEDING STARTS AFTER HEPARIN THERAPY

- Most common cause of drug related death in hospitalized patients = heparin
- Antidote = Protamine = 1 mg neutralizes 100 units of heparin; administer over 15 minutes. Empiric 25 to 50 mg IV [12, 6th ed., pg. 1017]

CAUSES OF DYSPNEA
Mnemonic: (SPACE) 2 [19, with modifications]

Dyspnea = a subjective shortness of breath with abnormal and uncomfortable awareness of breathing [10, pg. 1030]

S	**S**pontaneous pneumothorax	**S**hock ↑ hemodynamic stimulation
P	**P**neumonia	**P**ulmonary edema (CHF/ARDS)
A	**A**sthma	**A**telectasis
C	**C**OPD (bronchitis/emphysema)	**C**ardiac (CHF, AMI, valvular, AS, MS, AMR)
E	**E**mbolism-pulmonary (PE)	**E**ffusions (pulmonary)

PULMONARY

CAUSES OF PLEURAL EFFUSIONS
Mnemonic: B (CHAMPS) 3 [1, with modifications] [10, 5th ed., pg. 1005]

B	Boerhaave syndrome (left sided)

C	CHF Most common cause of effusions in Western countries	Cirrhosis (with ascites)	Chylothorax	
H	Hemothorax	Hypothyroidism	Hepatic infection (with upward spread)	
A	AFB positive = TB Most common cause of effusions developing countries	Asbestos	↓ Albumin	Aortic dissection (on left side)
M	Malignancy	METS	Meig's Syndrome	
P	Pneumonia "parapneumonic effusion"	PE (with infarct)	Pancreatitis (on left side)	
S	SLE (also RA) 12% pts SLE = Exud Effusion	Saline overload	Side effect of drugs (NSAIDS, etc.) (see below)	

CAUSES OF PLEURAL EFFUSIONS - SIDE EFFECTS OF DRUGS
Mnemonic: (MAP)

M	Macrodantin, sustained release = Macrobid (Nitrofurantoin)
A	Apresoline (Hydralazine)
P	Procainamide (Pronestyl, Procan)

All three drugs above (MAP), Dilantin or INH, can all cause **Lupus-Like Syndrome**
Amiodarone (Cordarone) causes pulmonary toxicity by pulmonary fibrosis

PULMONARY

PLEURAL EFFUSION PEARLS

Transudative pleural effusions
- Protein content < 3 gm/dl
- Pleural fluid vs. Serum protein RATIO < 3
- Pleural fluid vs. Serum LDH RATIO < 0.6
- LDH content less than 200

Transudative pleural effusions = CHF, cirrhosis, starvation (↓albumin), constrictive pericarditis, nephrotic syndrome and SVC obstruction

Exudative effusions have high amounts of protein and LDH; pH < 7.3
Examples = pneumonia, SLE, TB and malignancy

Exudates – **Ex**ceed (**high** amounts of protein and LDH effusion to serum ratios)
Effusion protein/serum protein > 0.5
Effusion LDH/serum LDH > 0.6

Pleural effusions: Need 200 cc of fluid to be seen on PA or lateral CXR
                    Need <50 cc of fluid if lateral decubitus film

CAUSES OF HEMOPTYSIS

Mnemonic: (BIC) 3 [1, with modifications]

	Common Causes	Less Common Causes	
B	Bronchitis	Bronchiectasis	Bechet and other Alveolar hemorrhage syndromes
I	Infection (TB, lung abscess, bronchitis, pneumonia)	Infarct (PE) Only 30% of patients with PE present with hemoptysis [10, 5th ed., pg. 1217]	Idiopathic Cryptogenic hemoptysis Common = 28%
C	Cancer	Cardiac (CHF, MS) CHF = 75% of cardiac cases	Cocaine "Crack lung"

Other causes
Alveolar hemorrhage syndromes: Bechet, Goodpasture syndrome and Wegener granulomatosis
Hematologic: platelet dysfunction, anticoagulant therapy and uremia (platelet dysfunction)
Cystic Fibrosis
Trauma

Epidemiology (in industrialized countries) [12, 6th ed., pg., 465]
Idiopathic = "cryptogenic hemoptysis" = 28%
Neoplasm = 28%
Infectious, non-TB = 25%
Miscellaneous = 13%
TB = 5%

Mild Hemoptysis < 20 ml in 24hrs; Moderate = 20 to 600ml; Severe = > 600 ml [12, 6th ed., pg., 465]

TB = cough is the most common presenting symptom, not hemoptysis; hemoptysis is usually minor in acute infection; later → major cause of massive hemoptysis; [10, 5th ed., pg. 1906]

PULMONARY

SOB is unusual; 15% get extrapulmonary manifestations
Aspergilloma as a superinfection with TB can also cause massive hemoptysis [10, 5th ed., pg., 1907]

HIV patient with hemoptysis = consider *Strep. pneumo.* or TB [10, 5th ed., pg., 1847]

Children with TB, most common CXR findings = hilar adenopathy, mediastinal lymphadenopathy or consolidated pneumonia

Evaluation [12, 6th ed., pg., 466]
CXR = abnormal in 70-85%; if neoplasm → abnormal in 80-90%
Massive bleeding = rigid bronchoscopy
Less severe bleeding = fiberoptic bronchoscopy
CT chest

COMMUNITY ACQUIRED PNEUMONIA (CAP)
[Clinical Infectious Diseases 2007;44:S27-S72]

The initial treatment of CAP is **empiric** and must cover these 6 bugs

"Typical" pathogens
- *Streptococcus pneumoniae*
- *Moraxella catarrhalis*
- *Haemophilus influenzae*, nontypeable

"Atypical" pathogens = not detectable on GS or cultivatable on standard bacteriologic media
- *Mycoplasma pneumoniae*
- *Chlamydophila pneumoniae*
- *Legionella*

MILD (AMBULATORY) CAP
[Clinical Infectious Diseases 2007;44:S27-S72]

The most common pathogens →
Strep. pneumoniae, M. pneumoniae, C. pneumoniae, and *H. influenzae*

- *Mycoplasma* = most common among patients <50 years of age without significant comorbid conditions or abnormal vital signs, whereas *Strep. pneumoniae* was the most common pathogen among older patients and among those with significant underlying disease.

- *Hemophilus* infection was found in 5%—mostly in patients with comorbidities

- The use of fluoroquinolones to treat ambulatory patients with CAP without comorbid conditions, risk factors for DRSP, or recent antimicrobial use is discouraged because of concern that widespread use may lead to the development of fluoroquinolone resistance.

COMMUNITY ACQUIRED PNEUMONIA (CAP)
[Clinical Infectious Diseases 2007;44:S27-S72]

- Blood cultures are low yield = 5 to 14%
- Blood cultures are *optional* for all hospitalized patients with CAP but should be performed selectively Get blood cultures if severe CAP, immunocompromised or if your yearly merit increase is affected by a hospital core measure list which includes blood cultures for CAP.
- There is strong evidence for the recommendation of **combination** empirical therapy for non-ICU and ICU

PULMONARY

CAP [Clinical Infectious Diseases 2007;44:S27-S72]

TREATMENT FOR NON-ICU – CAP

- Ceftriaxone *or* Cefotaxime *or* Unasyn *or* Ertapenem (Invanz) + Macrolide
- A respiratory fluoroquinolone can be used for penicillin-allergic patients in non-ICU CAP

TREATMENT FOR ICU – CAP
[Clinical Infectious Diseases 2007;44:S27-S72]

- Therapy with a respiratory fluoroquinolone alone is not established for severe ICU CAP
- Ceftriaxone *or* Cefotaxime *or* Unasyn **plus** either azithromycin (level II evidence) or a respiratory fluoroquinolone (level I evidence) (strong recommendation)
- For penicillin-allergic patients → respiratory fluoroquinolone + aztreonam

Pneumonia Severity Score for Elderly Patients	
CHARACTERISTIC	**POINTS**
Historical Findings	
Age - Men	Men (years) Women (years - 10)
Nursing home resident	10
Coexisting disease	
Neoplastic disease	30
Liver disease	20
Congestive heart failure	10
Cerebrovascular disease	10
Renal disease	10
Physical Examination Findings	
Altered mental status (acute)	20
Respiratory rate > 30	20
Systolic BP < 90 mmHg	20
Temperature < 35°C or > 40°C	15
Pulse > 125/min	10
Diagnostic Testing Findings	
Arterial pH < 7.35	30
BUN > 30 mg/dL	20
Sodium < 130 mmol/L	20
Glucose > 250 mg/dL	10
Hematocrit < 30%	10
PaO2 < 60 mmHg (or SaO2 < 90%)	10
Pleural effusion	10

Risk Class	Point Score	Mortality (%)	Disposition
2	< 70	0.6	Outpatient
3	71-90	2.8	Brief Inpatient
4	91-130	8.2	Inpatient
5	> 130	29.2	Inpatient

Fine and colleagues developed a *prognostic model*, the **Pneumonia Severity Index (PSI), for thirty day mortality** in patients with CAP. Patients are assigned to one of five risk classes (1=lowest risk and 5=highest risk) based upon a point system. Outpatient management is suggested for Class 1 and 2, brief inpatient for class 3 and traditional hospitalization for Classes 4 & 5. Severe pneumonia may require intensive care unit (ICU) admission. Guidelines should not supersede clinical judgment

SEVERITY OF ILLNESS SCORE
Mnemonic: (CURB 65) [Sanford, 2009, pg. 36; AnIM 118:384,2005]

C	Confusion	1 Point
U	Uremia - BUN > 19 mg/dl	1 Point
R	Respiratory rate - RR> 30/min	1 Point
B	Blood pressure, low - BP < 90/60	1 Point

65	Age > 65	1 Point

If score = 1, outpatient ok; If > 1 hospitalize. The higher the score the higher the mortality

IDSA / ATS GUIDELINES ICU ADMISSION DECISION
[Clinical Infectious Diseases 2007;44:S27-S72]

Major Criteria
- Mechanical ventilation or
- Shock (SBP < 90 mmHg) requiring pressors

Minor Criteria – presence of 3 criteria → admit ICU
- Respiratory rate > 30 breaths/min at admission
- Arterial oxygen pressure / fraction inspired oxygen (PaO2/FIO2) < 250 mmHg
- Multilobar infiltrates
- Confusion / disorientation
- BUN > 20 mg/dL
- Leukopenia from infection (< 4,000 cells/mm3)
- Thrombocytopenia (<100,000 cells/mm3)
- Hypothermia < 36°C
- Hypotension requiring aggressive fluid resuscitation

PULMONARY

EPIDEMIOLOGIC CONDITIONS AND/OR RISK FACTORS RELATED TO SPECIFIC PATHOGENS IN CAP
[Clinical Infectious Diseases 2007;44:S27-S72]

CONDITION	COMMONLY ENCOUNTERED PATHOGEN(S)
Alcoholism	*Streptococcus pneumoniae*, oral anaerobes, Acinetobacter species, *Mycobacterium tuberculosis*
COPD and/or smoking	*Haemophilus influenzae, Pseudomonas aeruginosa*, Legionella species, *S. pneumoniae, Moraxella cararhalis, Chlamydophila pneumoniae*
Aspiration	Gram-negative enteric pathogens, oral anaerobes
Lung abscess	CA-MRSA, oral anaerobes, endemic fungal pneumonia, M. tuberculosis, atypical mycobacteria
Exposure to bat or bird droppings	*Histoplasma capsulatum*
Exposure to birds	*Chlamydophila psittaci* (if poultry: avian influenza)
Exposure to rabbits	*Francisella tularensis*
Exposure to farm animals or parturient cats	*Coxiella burnetti* (Q fever)
HIV infection (early)	*S. pneumoniae, H. influenzae, M. tuberculosis*
HIV infection (late)	The pathogens listed for early infection plus Pneumocystis jirovecii, Cryptococcus, Histoplasma, Aspergillus, atypical mycobacteria (especially Mycobacterium kansasii), P. aeruginosa, H. influenzae
Hotel or cruise ship stay in previous 2 weeks	Legionella species
Travel to or residence in SW United States	Coccidioides species, Hantavirus
Travel to or residence in SE and East Asia	*Burkholderia pseudomallei*, avian influenza, SARS
Influenza active in community	Influenza, S. pneumo, Staph aureus, H. influenzae
Cough >2 weeks with whoop or posttussive V	*Bordetella pertussis*
Structural lung disease (e.g., bronchiectasis)	*Pseudomonas aeruginosa, Burkholderia cepacia, S. aureus*
Injection drug use	S. aureus, anaerobes, M. tuberculosis, S. pneumoniae
Endobronchial obstruction	Anaerobes, S. pneumoniae, H. influenzae, S. aureus
In context of bioterrorism	*Bacillus anthracis* (anthrax), *Yersinia pestis* (plague), *Francisella tularensis* (tularemia)

PULMONARY

PULMONARY PEARLS

- Adequate sputum specimen must have > 25 PMN's and < 10 squamous epithelial cells

- Most common cause of bacterial pneumonia = *Strep pneumoniae* √ rusty/blood sputum

- Most common non-bacterial pneumonia in adults < 50 y/o = *Mycoplasma pneumonia*

- A common cause of pneumonia in alcoholics = *Klebsiella pneumonia*, √ "currant jelly" sputum. *Strep pneumoniae* = most common cause of bacterial pneumonia in alcoholics [12, 6th ed., pg., 449]

- Neonatal pneumonia = GBS, *E. coli* and *Klebsiella*

- Pneumonia 1-3 months = Chlamydophila, CMV, Pneumocystis

- Most common cause of pneumonia in children < 6 months and 3-5 years old = RSV and parainfluenza Children > 5 y/o and young adults *Mycoplasma* is common

- Most common cause of viral pneumonia in adults = *Influenza*
  - *Strep pneumoniae* = most common cause of bacterial pneumonia .... However → also consider *Staph* pneumonia during outbreaks of influenza.
  - *Staph* pneumonia = necrotizing with cavitations and pneumatically formation [10, 5th ed., pg., 987]

- The majority of deaths from viral pneumonia occur during winter months

- Most common source of infection in **septic** patient = respiratory system [10, 5th ed., pg., 1961]

- Organisms that cause pneumonias which present with fever and relative bradycardia = Legionella, Q fever, Psittacosis and Tularemia (note: all 4 also ↑LFTs) [12, 6th ed., pg., 971]

- Most common pathogens to generate cavitary lesions = TB, anaerobic bacteria, aerobic gram-negative bacilli, *S. aureus,* and fungal disease [10, 5th ed., pgs 990-993]

- Most common complication of cavitary TB = endobronchial spread (seen on x-ray as 5 to 10 mm poorly defined nodules clustered in dependent portions of the lungs [10, 5th ed., pg. 1907]

- TB = world's leading cause of infectious death; 1/3 of world population is infected [10, 5th ed., pg 1903]

- PPD often negative in late HIV; minimum triple drug therapy for TB, if HIV + TB four drugs [10, 5th ed., pg 1848]

- Carcinoma suspected on CXR → remember L.A. (Los Angeles) is located on the periphery of the US → Large cell and Adenocarcinoma present as peripheral masses on CXR; Squamous cell and Small cell (most malignant) present as central or hilar tumors.

- Most common cause of cancer deaths in males and females = bronchogenic CA

- Paraneoplastic syndromes = ↓Na+/ SIADH, ↑Ca2+/ parathormone, ↓Ca2+/ Calcitonin, ↑ACTH/ Cushing syndrome, gynecomastia/ gonadotropins, ↑Serotonin/ Carcinoid syndrome)

PULMONARY

- CXR findings of COPD = flat diaphragm, ↑AP diameter, enlarged retrosternal space, blebs

- If you suspect chest pain is due to pneumothorax, ask for inspiratory and **expiratory** views on chest film

- Albuterol β 2 agonist ↑ **cAMP** → bronchodilation

- Atrovent is an anticholinergic agent that ↓ **cGMP** → bronchodilation

- Epi 1:1000 IM in the thigh (vastus lateralis) results in higher and more rapid maximum plasma concentrations

- Most common occupational lung disease worldwide = Silicosis (found in sand, granite, sandstone, flint, slate, and in coal and metallic ores. Cutting, breaking, crushing, drilling, grinding, or abrasive blasting of these materials may produce fine silica dust *emedicine.medscape.com/article/302027-overview*
    - Fever, SOB, CP, Cor pulmonale and **cyanosis**
    - Susceptible to TB (silicotuberculosis) suspect silica damages pulmonary macrophages, inhibiting their ability to kill mycobacteria.
    - Silicosis is an irreversible condition with no cure. Treatment = Lung transplantation

PULMONARY

RENAL-UROLOGY

CAUSES OF HEMATURIA
Mnemonics: (IN STITCHES2) [ANK]

I	Infection (pyelonephritis, hemorrhagic cystitis, prostatitis, TB, Schistosomiasis)
N	Necrosis = Papillary Necrosis (NSAIDs, DM, sickle cell disease)

S	Stones, Kidney Stricture of the urethra
T	Toxins (direct kidney injury or rhabdo) Toxemia of pregnancy
I	Intrinsic Kidney Disease (glomerulonephritis, PKD, Medullary Sponge, IgA nephropathy (Berger disease), hereditary nephritis (Alport syndrome) Iatrogenic (post procedure)
T	Trauma Thrombosis of renal vein
C	Coagulopathy (Hemophilia, ↑coumadin or heparin toxicity) Cancer (renal, bladder, prostate, Wilms' tumor/nephroblastoma)
H	Hemoglobinopathy (sickle cell disease) Huge prostate (BPH)
E	Endocarditis Expanding AAA may erode into urogenital tract; also Aortic Dissection
S	SLE (50% nephritis) other immunologic disease = (ITP, HSP (IgA complexes), Goodpasture syndrome, Wegener granulomatosis, polyarteritis nodosa (PAN); nonimmune causes = TTP, HUS Serum sickness

Don't confuse Berger with Buergers's disease (thromboangiitis obliterans) = recurring inflammation and thrombosis of small and medium arteries and veins of the hands and feet. Strongly associated with smoking

Berger, Buergers, Bugger, Burger say that fast 20x

Triad of sinusitis, pulmonary infiltrates, and nephritis = Wegener granulomatosis

RENAL \ UROLOGY

KIDNEY STONES

3 x more common males > females

Calcium oxalate or Ca phosphate stones
- Most common 75%; radio-opaque
- Causes
  - Hypercalciuria → GI = reabsorption; Bone = resorption (primary hyperPTH); Kidney = Ca^{2+} leak
  - Idiopathic, sarcoid, hyper-thyroidism,
  - Paget's disease, increased incidence from genetic predisposition, PUD taking antacids, IBD

Magnesium-ammonium-phosphate (Staghorn or struvite stones)
- 15%, associated with alkaline urine (pH > 7.6),
- secondary to recurrent infections: Proteus, Pseudomonas, Klebsiella
- Urea → split to NH_3 by Urease → binds H^+ → NH_4^+ [10, 5th ed., pg. 1415]

Uric acid
- 10%, radiolucent, associated with acidic urine; 25% patients with gout get stones

Cystine stones
- 1%, 2 0 inborn error in metabolism → ↑secretion of cystine → form staghorns

Medications That Can Cause Stones
- Indinavir (Crixivan) - radiolucent stones; HIV med, protease inhibitor

Kidney Stone Pearls
90% stones < 4 mm pass spontaneously
50% stones 4 to 6mm pass spontaneously
10% stones > 6mm pass spontaneously [12, 6th ed., pg. 621]

- Diagnostic test = non-contrast spiral CT – may find alternate diagnosis …. AAA or appy consider Ultrasound to evaluate for hydronephrosis/ hydroureter in patients with known history of kidney stones
- Most common location of stone at first diagnosis = distal ureter 70%
- Most common site = ureterovesical junction (UVJ)
- Treatment: Fluid, NSAIDs decrease ureteral spasm, narcotics and Tamsulosin (Flomax)

NEPHRITIC SYNDROME
Mnemonics: (PHARAOH) [ANK, Mnemonic provided by Dr. Justino Dalio]

P	**P**roteinuria
H	**H**ematuria
A	**A**zotemia
R	**R**BC casts
A	Immunoglobulin **A** (IgA) nephropathy glomerulonephritis (Berger disease) = Most common cause of glomerulonephritis worldwide
O	**O**liguria
H	**H**TN

RENAL \ UROLOGY

TESTICULAR TORSION

- Absent cremasteric reflex
- Pain may be constant, intermittent but is *not* positional
- Manually detorsion of testicle → stand at patient's feet → detorsion in similar fashion to opening a book → patient's right testicle → rotated counterclockwise; left testis → clockwise
- Manually detorsion is not curative, attempt while waiting for surgery [10, 5th ed., pg. 1423]
- If you cannot exclude torsion by H&P and imaging studies → surgery for scrotal exploration, *regardless* of time since onset of symptoms [12, 6th ed., pg. 617]
- "Blue-dot sign" on testicular exam = Torsion of the testicular appendix

PENILE FRACTURE

- Traumatic rupture of the tunica albuginea and corpus cavernosum urologic emergency.

- Sudden blunt trauma of the penis in an erect state causes rupture in one or both corpora.

- Concomitant urethral injury may also occur.

- Clinical diagnosis: Patients usually hear a "popping" or "cracking" sound, followed by significant pain, immediate flaccidity and skin hematoma/ecchymosis.

- Conservative management has fallen out of favor because of high complication rates.

- Surgical repair is necessary to expedite relief of pain and to prevent potential complications → surgical evacuation of hematoma and suture apposition of the disrupted tunica albuginea [12, 6th ed., pg. 617]

- Complications of conservative management include: missed urethral injury, penile abscess, nodule formation at the site of rupture, permanent penile curvature, painful erection, painful coitus, erectile dysfunction, corporal urethral fistula, arterial venous fistula and fibrotic plaque formation.

PRIAPISM

Engorgement of the corpora cavernosa. Ventral corpora spongiosum and glans penis usually remain flaccid

Two types
- **Low-flow (ischemic) priapism** (most common)
    - Veno-occlusion; painful
    - Most common cause = Sickle cell anemia

- **Arterial (non-ischemic) high-flow** priapism
    - Secondary to a rupture of a cavernous artery.
    - Rare; usually not painful; Causes = penetrating penile trauma or a blunt perineal injury

Other Causes of Low-flow priapism

- **Medications**

RENAL \ UROLOGY

- Injectable medications to induce an erection = papaverine, phentolamine (non-selective alpha-1, alpha-2 adrenergic-receptor **antagonist**), and prostaglandin E1
- Psychotropic medications = chlorpromazine, trazodone, quetiapine, thioridazine and citalopram (SSRI)
- Rebound hypercoagulable state with anticoagulants = heparin and warfarin
- Hydralazine, metoclopramide, omeprazole, hydroxyzine (atarax; vistaril), prazosin, tamoxifen, and androstenedione
- Cocaine, marijuana, ecstasy, and ethanol

- **Neurologic**
  - Spinal cord injury and anesthesia
  - Cauda equina syndrome

- **Neoplastic**
  - Primary or metastatic; leukemia and multiple myeloma

- **Infection**
  - Mycoplasma pneumoniae (Mechanism is thought to be a hypercoagulable state induced)
  - Malaria

Treatment Low-flow (vaso-occlusive) priapism [12, 6th ed., pg. 616]
- Terbutaline 0.25 to 0.5 mg SQ in deltoid, may be repeated in 20 to 30 min
- Oral pseudoephedrine, 60-120 mg orally due to its alpha-agonist effect (sympathomimetic amine → ↑endogenous norepinephrine from storage vesicles in presynaptic neurons)
- Next line of therapy = aspiration of corpus cavernosum first and injection of alpha-1-adrenergic agent phenylephrine (Neo-Synephrine)
- In adults, phenylephrine should be diluted with normal saline to provide a final concentration of approximately 100 mcg to 500 mcg per mL [http://www.uptodate.com/contents/priapism]
- Phenylephrine (Neo-Synephrine) 10mg/mL which = 10,000 mcg/mL
- Dilute 10mg/mL Phenylephrine in 100mL of normal saline =
- 100 mcg/mL of Phenylephrine (Neo-Synephrine) solution ---> draw up 1 to 5 mL to get recommended treatment dose of 100 mcg to 500 mcg per mL

Treatment High flow
- Observation alone may be sufficient as erectile function is usually unimpaired.
- Compression therapy may be successful in certain cases, especially children

UROLOGY PEARLS - TRAUMA

- Most common urologic injury = renal; 80% have concurrent injuries

- Most common renal injury = contusion 90% (Grade I)

- Renal injuries are Graded I to V; order CT with IV contrast

  Surgery: Uncontrolled hemorrhage, penetrating injuries, avulsed major renal vessel, extensive urine extravasation, Grade V injury (shattered kidney with avulsed hilum)

- Most common urethral injury = posterior (90%), above the urogenital diaphragm, associated with pelvic fractures (high riding or absent prostate on rectal exam)

RENAL \ UROLOGY

- Anterior urethral injury = 10%, direct trauma (kicks or straddle injuries)

- Blood at meatus = urethral trauma, no foley without first performing retrograde urethrogram

- Retrograde urethrogram = Toomey syringe into urethral meatus → inject 60 ml contrast over 30 to 60 seconds → xray during the last 10ml contrast injection

- Retrograde cystogram / retrograde CT cystography = after normal retrograde urethrogram → allow 300 to 400 ml contrast to flow by gravity from a Toomey syringe through a Foley catheter into bladder → clamp foley → AP and lateral films or CT of bladder taken → unclamp Foley → postevacuation film / CT

- Bladder injuries
    - Contusion = most common bladder injury → incomplete tear of bladder mucosa; large hematomas can alter bladder shape on cystogram = "pear-shaped" bladder
    - **Extraperitoneal** bladder rupture → heal spontaneously in 14 days with Foley
    - Most common type of bladder rupture 80% (rupture usually at bladder neck)
    - **Intraperitoneal** bladder rupture (injury is in the dome posteriorly, the only portion of dome covered by the peritoneum) → surgical repair [12, 6th ed., pg. 1626]

UROLOGY PEARLS

- Most common cause of hematuria in females = infection

- Most common cause of hematuria in men = BPH → urinary retention

- Most common cause of hematuria worldwide = Schistosomiasis → *Schistosoma haematobium* (blood fluke) *S. mansoni & S. japonicum* = esophageal varices; Dx = eosinophilia and identification of eggs in first-morning urine or stool or biopsy; Treatment = Praziquantel

- Degree of hematuria after blunt trauma does not correspond to degree of injury [12, 6th ed., pg. 1622]

- Pseudo-hematuria = blood tinged but *no* RBC's on UA consider myoglobinuria (from trauma, seizures, burns, sepsis – any cause of rhabdo), pyridium, rifampin, porphyria

- Spontaneous complete drainage of distended bladder ok, no need to clamp. Transient gross hematuria or hypotension is usually insignificant [12, 6th ed., pg. 620]

- If urinary retention is chronic or insidious → postobsturctive diuresis; observe 4 to 6 hours

- Fournier's gangrene = surgical debridement, fluid and antibiotic cover aerobic/anaerobic per [Sanford, 2009, pg. 52]

- Acute hydrocele is most likely associated with = testicular cancer (7-25% of patients)

- Acute Right-sided varicocele is most likely associated with = IVC thrombosis or compression of IVC by tumors

- Acute Left-sided varicocele is most likely associated with = Renal cell CA or obstruction of left renal vein

RENAL \ UROLOGY

Paraphimosis inability to reduce the retracted foreskin over the glans; ice, manual decompression; dorsal slit
Phimosis inability to retract the foreskin to visualize the glans; topical steroids 4-6 weeks; circumcision
Balanitis inflammation of the glans penis; Candida 40%, *Group B Strep.*, Gardnerella and anaerobes; occurs in 1/4 male sex partners of women infected with candida
Treatment = Fluconazole 150 mg po x 1 [Sanford, 2009, pg. 24] or antifungal creams [12, 6th ed., pg. 615]
Posthitis inflammation of the foreskin; consider adding 1st generation cephalosporin for bacterial infection
Balanoposthitis is inflammation of both the glans penis and surrounding foreskin
Recurrent Balanoposthitis can be the sole presenting symptom of = diabetes mellitus

UROLOGY PEARLS – STDS
[CDC STD UPDATE 2010]

Epididymitis [Sanford, 2009, pg. 24]
- < 35 y/o GC (ceftriaxone 250 mg IM) and Chlamydia (doxy 100 mg po bid x 10 day *not* zithro x1)
- > 35 y/o *E. coli* (quinolone x 10 to 14 days)
- Prehn's sign = scrotal elevation → ↓pain in epididymitis and not in torsion; know sign for test then forget it, since insensitive to distinguish from epididymitis from testicular torsion

Chancroid
- *Haemophilus ducreyi*, gram-negative bacillus = painful chancres (*Do Cry with H. ducreyi*)
- High rates of HIV infection among patient who have chancroid
- 10% of persons who have chancroid acquired in the US are co-infected with *T. pallidum* or HSV
- Treatment
  - Azithromycin 1 gm orally or
  - Ceftriaxone 250 mg IM in a single dose or
  - Ciprofloxacin 500 mg twice daily x 3 days

Granuloma Inguinale (Donovanosis)
- Calymmatobacterium granulomatis, gram-negative pleomorphic bacillus
- Beefy-red, velvety, painless ulcer with rolled border
- Diagnosis = organism within macrophages (Donovan bodies) in biopsy specimens taken from the advancing edge of the ulceration. Macrophages engulf clusters of organisms that look like microscopic safety pins = Donovan bodies
- Treatment = Doxy 100 mg po bid x 3 weeks minimum

Lymphogranuloma Venereum
- Chlamydia trachomatis serotypes L1-3
- Painless papules, vesicles, or ulcers
- Typically unilateral, painful inguinal lymphadenopathy ("groove" sign)
- Treatment = Doxy 100 mg po bid or Erythromycin base 500 mg four times daily x 3 weeks

Syphilis is Divided into Clinical Stages

Primary	Painless chancre
Secondary	9-90 d, after the onset of the chancre; maculopapular rash, + palms/soles
Latent	Asymptomatic
Late	Manifestations 10-20 yrs following primary infection CNS, CV, skin, and/or bone may be involved

RENAL \ UROLOGY

- **Diagnosis** - corkscrew-like spirochetes under dark field microscopy; serologic tests (VDRL, RPR) = non-specific
- **Treatment** - Primary, Secondary, Early Latent = Benzathine Penicillin G, 2.4 million units IM

- Jarisch-Herxheimer reaction = acute onset of fever, rigors, and possibly hypotension may occur within 24 hours of initiating treatment
- Condyloma lata = raised, flat, grayish papular lesions which are found in moist areas of the body including the anus, vulva, and scrotum; seen in secondary syphilis

Trichomoniasis
- Parasite *Trichomonas vaginalis*
- Treatment = Metronidazole 2 gm orally in a single dose
- Diagnosis = saline microscopy of vaginal secretions on wet slide prep (motile, flagellated trichomonads)
- Sensitivity = 60-70%
- Immunochromatographic capillary flow dipstick technology, and a nucleic acid probe test performed on vaginal secretions; more sensitive than vaginal wet preparation, however ↑false positives
- Culture if need definitive diagnosis

Bacterial Vaginosis (BV)
- Polymicrobial infection
- Clue cells on saline microscopy (bacteria adhered to vaginal epithelial cells)
- Metronidazole 500 mg twice daily for 7 days
- All symptomatic pregnant women with BV should be treated, regardless of trimester
- Woman's response to therapy and the likelihood of relapse or recurrence **not** affected by treatment of sex partner

RENAL \ UROLOGY

TOXICOLOGY

MNEMONICS AND PEARLS HANDBOOK

DIALYSIS CRITERIA
Mnemonic: (AEIOU) [ANK]
Note: Mnemonic AEIOU also used in "Causes of Coma"

A	**A**cidosis
E	**E**lectrolyte abnormalities (↑K+)
I	**I**ngestion of toxins (**I**sopropanol, **B**arbiturates, **A**mphetamines, **I**NH, **L**ithium, **T**heophylline, **E**thylene glycol, **A**SA, **m**ethanol)
O	**O**verload-fluid
U	**U**remic Pericarditis

INDICATIONS FOR DIALYSIS OF TOXINS
Mnemonic: (I BAIL TEAM)
Dialysis can BAIL your medical TEAM out of a life threatening situation

I	**I**sopropanol (isopropyl alcohol, rubbing alcohol; note: normal AG ↑ osmolar gap) [12, 4th ed. pg 769]

B	**B**arbiturates → long acting (charcoal hemoperfusion); Beta blockers (water soluble = atenolol)
A	**A**mphetamines
I	**I**NH
L	**L**ithium (level > 4 mEq/L or 2-4 if poor clinical condition)

T	**T**heophylline (charcoal hemoperfusion)
E	**E**thylene glycol (levels > 20 mg/dL; nephrotoxicity; metabolic acidosis is present; Triad = history, clinical presentation & lab results consistent with Ethylene glycol poisoning)
A	**A**SA (if seizures, CNS alteration, acidosis, serum levels > 100mg/dl)
M	**M**ethanol (visual or CNS dysfunction; methanol levels > 20 mg/dL; ingestion > 30 ml; pH < 7.15) **M**ushrooms = Amanita (charcoal hemoperfusion)

Note: dialysis has NOT been shown to be effective in Acetaminophen, BZP, Clonidine, Digoxin, Dilantin [12, pg. 810], MAOI OD, Heroin (↑ Vd), Organophosphate poisonings [12, pg. 826] TCA OD-highly bound and high volume of distribution; Iron [12, 6th ed., 1123]

Drugs which can be dialyzed have:
- Small molecular weight
- Small volume of distribution
- Water Solubility
- Lack protein binding

TOXICOLOGY

OVERDOSES WHERE BICARB MAY BE A TREATMENT OPTION
Mnemonic: (LCD BITS)

L	**L**amictal (lamotrigine)
C	**C**ocaine overdose: treat if wide-complex tachycardia [10, 4th ed. pg. 1436], [Ref 29, 2010 AHA pg. 57]
D	**D**iphenhydramine **D**arvon = propoxyphene (similar to type IA antidysrhythmic agents)

B	**B**arbiturates - Long acting only, eg phenobarbital and barbital; note: butalbital found in Fioricet is short-acting, 25% of phenobarbital is excreted in urine; alkaline urine pH → ↑ excretion
I	**I**NH (Isoniazid); treatment = pyridoxine (B6)
T	**T**CAs: treat if wide-complex tachycardia, hypotension and ventricular arrhythmias
S	**S**alicylates: favors the formation of ionized salicylate → ↓ ASA reabsorption → ↑ exertion

CHARCOAL INEFFECTIVE
Mnemonic: (MP LICE)

M	**M**etals-heavy (Treatment = whole bowel irrigation = PEG (polyethylene glycol)
P	**P**etroleum distillates (Hydrocarbons)

L	**L**ithium
I	**I**ron
C	**C**austics
E	**E**thylene glycol (Also → ethanol, methanol and isopropyl alcohol)

WHOLE BOWEL IRRIGATION
Mnemonic: (SLIMS)

S	**S**ustained release
L	**L**ithium
I	**I**ron
M	**M**etals (heavy)
S	**S**tuffers

TOXICOLOGY

RADIOPAQUE SUBSTANCES
Mnemonic: (BET A CHIP) [ANK/13]

B	**B**arium
E	**E**nteric coated ASA
T	**T**CA's

| A | **A**ntihistamines |

C	**C**hloral hydrate
H	**H**eavy metals (Iron, lead, etc. Note: MVI with Iron-not seen on x-ray)
I	**I**odine
P	**P**henothiazine → *Examples*: chlorpromazine (Thorazine), prochlorperazine (Compazine), mesoridazine (Serentil), thioridazine (Mellaril) and promethazine (Phenergan)

CAUSES OF NON-ANION GAP ACIDOSIS
Mnemonic: (HARD CUP) [ANK]

H	**H**yperalimentation
A	**A**cetazolamide (Diamox)
R	**R**TA (proximal)
D	**D**iarrhea

C	**C**holestyramine
U	**U**terosigmoidostomy
P	**P**ancreatic fistulas

(Causes of ↓ AG = Acetazolamide, Ammonium Cl, Bromide, Iodide, Lithium, Polymyxin B, Spironolactone, Sulindac)

CAUSES OF ANION GAP ACIDOSIS
Mnemonic: (CAT MUDPILES) [ANK]

C	**C**O, CN (inhibit cytochrome oxidase a-a3 → ↑ lactate)
A	**A**lcoholic ketoacidosis
T	**T**oluene (secondary) to acidic metabolites

M	**M**ethanol, Metformin
U	**U**remia
D	**D**KA
P	**P**araldehyde
I	**I**NH (Isoniazid, inhibits lactate ↔ pyruvate, therefore → ↑ lactate) Iron (hypovolemia and anemia → tissue hypoperfusion → ↑ lactate)
L	**L**actic acidosis
E	**E**thylene glycol
S	**S**alicylates

TOXICOLOGY

Salicylates
- Inhibit Kreb's cycle dehydrogenases → ↑ lactate. Also, uncouple oxidative phosphorylation → ↑ heat (fever) and ↑glycolysis → hypoglycemia → ↑lipid metabolism → ↑ketones → ↑ acidosis

ETOH
- Normal AG and ↑ osmolal gap; however if ETOH ketoacidosis, AG is↑

Causes of Lactic Acidosis
- CO, CN, ASA, INH, Fe toxicity, ETOH abuse, hypoxemia, metformin, ritodrine, seizure and shock [9, pg. 199]

METHANOL AND ETHYLENE GLYCOL
- Both inhibit mitochondrial respiration → ↓ intracellular NAD/NADH ratio →
- ↑ anaerobic glycolysis → ↑ lactate

Methanol (windshield washer fluid)
- →ADH formaldehyde → formic acid → accounts for most of the AG metabolic acidosis and ocular toxicity; symptoms may be delayed for up to 12 to 18 hrs

Methanol Treatment
- ETOH or Fomepizole (4-methylpyrazole)
- Folate 50 mg IV q 4 hours for several days
- Bicarb for severe acidosis
- Fomepizole →
    - 15 mg/kg over 30 minutes
    - 10 mg/kg q 12 hours x 4 doses
    - More frequent dosing required during dialysis (Fomepizole is removed with dialysis)

Methanol Indications for dialysis [12, 6th ed., pg. 1069]
- Visual or CNS dysfunction
- Peak methanol levels > 20 mg/dL
- Ingestion > 30 ml
- pH < 7.15

Ethylene Glycol
- → ADH Glycoaldehyde → Glycolic acid (main player for ↑AG metabolic acidosis)
- → → Formic acid and Oxalic acid; Oxalic acid → binds Ca 2+ → urine crystals
- Used as de-icer, antifreeze and coolant

Ethylene Glycol 3 Phases of Toxicity
1) CNS = 1 to 12 hrs; ataxia, slurred speech, hallucinations, seizures, nystagmus, coma, death
2) Cardiopulmonary=12 to 24 hrs; ↑HR, ↑BP, ↑RR = most common; CHF, ARDS, CV collapse
3) Nephrotoxic = 24 to 72 hrs; flank pain, ATN, Ca Oxalate crystals → ↓Ca (hypocalcemia in EG toxicity is a common test question)

Ethylene Glycol Treatment
- ETOH or Fomepizole
- Thiamine (B1) 100 mg and Pyridoxine (B6) 100 mg IM or IV (cofactors needed in the metabolic pathway of ethylene glycol for the conversion to → nontoxic compounds)
- Replace Ca 2+ if necessary; 1 amp Ca Gluconate
- Bicarb for severe acidosis [12, 6th ed., pg. 1068]

TOXICOLOGY

Ethylene Glycol Indications for dialysis [12, 6th ed., pg. 1070]
- Triad = history, clinical presentation & lab results consistent with Ethylene glycol poisoning
- Ethylene glycol levels > 20 mg/dL
- Signs of nephrotoxicity
- Metabolic acidosis is present

SUBSTANCES CAUSING AN OSMOLAR GAP

Mnemonic: (I MADE GAS)
Note: You would too, if you ate GYROS everyday!

I	Isopropyl alcohol

M	Methanol / Mannitol
A	Alcohol (ETOH)
D	DKA (due to acetone)
E	Ethylene Glycol

G	Glycerol
A	Acetone
S	Sorbitol

RESPIRATORY COMPENSATION FOR METABOLIC ACIDOSIS
$PCO_2 = 1.5 [HCO_3] + 8 \pm 2$

Anion Gap = Na – (CL + HCO3)	Normal = 12 ± 2
Osmolarity = 2 [Na] + glucose/18 + BUN/2.8	Normal = 280-290
Osmolal Gap = Lab determined osmolarity – calculated osmolarity	Normal = 5-10

Note: every 4.2mg/dl of alcohol = 1 milliosmol; therefore if Osmolal Gap is elevated, order serum ethanol level and make correction; if Osmolal Gap still elevated think of the mnemonic, "I MADE GAS"

FACTORS INCREASING THEOPHYLLINE HALF-LIFE
Mnemonic: (COP CELICA) Imagine a COP chasing a Toyota CELICA [4, with modifications and 12, pg. 796]

C	Contraceptive-oral
O	Obesity
P	Pregnancy, 3rd trimester reduces the clearance of theophylline

C	Cipro
E	Erythromycin / Elderly and Neonates
L	Liver Disease
I	Inderal (Propranolol)
C	Cimetidine (Tagamet)
A	Allopurinol

TOXICOLOGY

FACTORS DECREASING THEOPHYLLINE HALF-LIFE
Mnemonic: (SMOKING CPR)

If you smoke, you may need CPR

SMOKING	
C	**C**arbamazepine (Tegretol)
P	**P**henytoin and **P**henobarbital
R	**R**ifampin

DRUGS THAT INCREASE DIGOXIN LEVELS
Mnemonic: (VAN – PQ)

V	**V**erapamil (Isoptin, Calan, Sustained release = Isoptin SR, Calan SR, Verelan, Covera)
A	**A**miodarone (Cordarone)
N	**N**ifedipine (Procardia, Adalat)

P	**P**rozac
Q	**Q**uinidine

POISONING ASSOCIATED WITH FEVER
Mnemonic: (SAL$_2$T$_3$ ASAP)

S	**S**alicylates		
A	**A**mphetamines		
L	**L**SD (rare) [12, pg. 781]	**L**ithium	
T	**T**CA's	**T**heophylline	**T**hyroxine

A	**A**nticholinergics (antihistamines, phenothiazine, TCA's)
S	**S**ympathomimetics (cocaine, amphetamines)
A	**A**ntihistamines
P	**P**CP (may be hypo-or hyperthermic)
MAO Inhibitor overdose	

POISONING ASSOCIATED WITH HYPO-THERMIA
Mnemonic: (COOLS) [ANK]

C	**C**arbon monoxide
O	**O**piates
O	**O**ral hypoglycemics, insulin
L	**L**iquor
S	**S**edative hypnotics (barbiturates, benzodiazepines, chloral hydrate, etc.)

TOXICOLOGY

DIAPHORETIC SKIN
Mnemonic: (SOAP)

S	**S**ympathomimetics
O	**O**rganophosphates
A	**A**SA (salicylates)
P	**P**CP

SIGNS AND SYMPTOMS OF CHOLINERGIC EXCESS
Mnemonic: (SLUDGE BAM) [ANK]

S	**S**alivation
L	**L**acrimation
U	**U**rination
D	**D**iaphoresis
G	**G**I – motility (diarrhea)
E	**E**mesis

B	1) **B**radycardia 2) **B**ronchoconstriction → **B**ronchospasm 3) **B**ronchorrhea
A	**A**bdominal cramps
M	1) **M**iosis 2) **M**uscle cramps, weakness and fasciculations

- Acetylcholinesterase (ACHE) inhibitors → cholinergic excess.

- Examples of acetylcholinesterase inhibitors include organophosphate and carbamate insecticides, nerve gases (sarin, soman, tabun and VX) and therapeutic agents (edrophonium/tensilon, pyridostigmine, physostigmine, and neostigmine)

- **Organophosphate poisoning:** the insecticides' phosphate radicals covalently bind to active serine sites on ACHE → enzymatically inert proteins; **irreversible** inhibition. Examples include: isoflurophate, echothiophate, pesticides such as malathion and parathion and toxins such as sarin and other nerve gases [12, pg. 823;10, 4th ed. pg. 1403]

- **Carbamates** → carbamylation of ACHE → **reversible** inhibition, because of the bond spontaneously breaks within 4 to 8 hours with regeneration on ACHE to the active form. Examples include: pyridostigmine, physostigmine, neostigmine and pesticides like carbaryl

TREATMENT CHOLINERGIC EXCESS

1) **ABC's**
2) Decontamination, lavage and charcoal 1 gm/kg
3) **Atropine** 3 mg IV, (peds 0.05 mg/kg IV), large amounts (10-20 mg) may be needed over 24 hours. Atropine blocks the action of acetylcholine at *muscarinic* (**SLUDGE BA** presentation), not nicotine receptors responsible for **M**uscle cramps, weakness and fasciculation and tachycardia

TOXICOLOGY

4) **Pralidoxime (2-PAM)** adults 1-2 gm IV, (peds 23-30 mg/kg over 3-5 minutes) [9, pg. 355] 2-PAM reverses the cholinergic *nicotinic* effects that are unaffected by atropine alone.
Use in carbamate poisoning is controversial [12, 5th ed., pg. 2189]
Mechanism:
   - Reactivation of cholinesterase by cleaving the phosphorylated active sites
   - Direct detoxification of unbound organophosphate and
   - Endogenous ANTI-cholinergic effect [12, pg. 826]

SIGNS AND SYMPTOMS OF ANTI-CHOLINERGIC TOXICITY
[ANK]

- Dry as a bone (dry skin first symptom)
- Hot as Hades (hyperthermia)
- Blind as a bat (mydriasis)
- Mad as a hatter (delirium, hallucinations)
- Red as a beet (flushing with hyperthermia)

Tachycardia, Urinary Retention, Hypoactive or absent bowel sounds

ANTI-CHOLINERGIC TOXICITY EXAMPLES

- Anti-parkinsonian drugs, anti-histamines (H1-receptor blockers), phenothiazine, some mushrooms, TCAs, belladonna alkaloids (**jimson weed**, atropine, scopolamine), Atrovent and chemically related drugs to TCA's such as carbamazepine (Tegretol), cyclobenzaprine (Flexeril)
- Note: sympathomimetics present with similar symptoms as anti-cholinergic excess .... however, **sympathomimetics → sweating and + bowel sounds**
- Treatment: Physostigmine 1-2 mg IVP slowly over 5"
- If give too fast → SLUDGE BAM or seizures
- Also, most common side effect of physostigmine = seizures; other side effects =↓HR,↓BP and SLUDGE BAM
- t ½=30-60 minutes, may repeat in 20 minutes if no effect

SUBSTANCE CAUSING NYSTAGMUS
Mnemonic: (PCP To PETS MEALS)
If you want to see your pets go crazy and demonstrate nystagmus, add PCP to your pets' meals

PCP	PCP (Phencyclidine)
To	Thiamine depletion

P	Phenytoin
E	ETOH
T	1) Tegretol 2) Thiamine depletion
S	Solvents

M	Methanol
E	Ethylene glycol
A	Alcohols (isopropyl alcohol)
L	Lithium [10, 4th ed. pg. 1392]
S	Sedative hypnotics

TOXICOLOGY

MIOSIS
Mnemonic: (COPS) 2 [ANK]

C	**C**lonidine	**C**holinergics
O	**O**piates	**O**rganophosphates
P	**P**henothiazine	**P**ilocarpine
S	**S**edative hypnotics	Stroke (pontine bleed)

PCP Intoxication (cholinergic) = **miosis** (2.1%) or (anticholinergic findings) = **mydriasis** (6.2%)
Horner syndrome = miosis

MYDRIASIS
Mnemonic: (4 - AAAA) [ANK]

A	**A**ntihistamines
A	**A**ntidepressants (TCAs)
A	**ANTI**-cholinergics
A	**A**mphetamines & other Sympathomimetics (cocaine)

CAUSES OF SEIZURES
Mnemonic: (U HIT)2 (OTIS CAMPBELL) 3 [ANK/13 with modifications]

MEDICAL			
U	**U**remia	**U**sed up oxygen (hypoxia)	
H	**H**ypo's (Na, Ca, Mg)	**H**ypo-glycemia	
I	**I**nfection	**I**diopathic	
T	**T**rauma (SAH, etc.)	**T**umor (brain)	

TOXICOLOGY - Otis Campbell = The "town drunk" on "The Andy Griffith Show"

O	**O**rganophosphates	**O**piates	**O**lanzapine (Zyprexa)
T	**T**CA's	**T**heophylline	**T**ramadol (Ultram)
I	**I**NH	**I**ron	**I**nhalants
S	**S**ympathomimetics	**S**alicylates	**S**SRIs

C	**C**ocaine	**C**O	**C**N and **C**amphor
A	**A**mphetamines	**A**nticholinergics	**A**nesthetics - local
M	**M**DMA	**M**eperidine (Demerol)	**M**ushrooms (Gyromita esculenta)
P	**P**CP	**P**hysostigmine	**P**ropoxyphene (Darvon)
B	**B**-blockers	**B**enadryl	**B**upropion (Wellbutrin, Zyban)
E	**E**TOH withdrawal	**E**tomidate	**E**phedra
L	**L**ithium	**L**SD [12, pg. 781]	**L**exapro (escitalopram)
L	**L**ead	**L**idocaine	**L**indane

TOXICOLOGY

SEIZURE HISTORY
Mnemonic (B COLD)

B	**B**ladder or Bowel incontinence

C	**C**haracter (type of seizure)
O	**O**nset (when did it start, what was the patient doing)
L	**L**ocation (where did the activity start)
D	**D**uration (how long did it last)

Also, previous episodes of seizure activity, previous medical history, meds and trauma history.

MORE CAUSES OF SEIZURES

Treat INH (isoniazid) seizures with IV pyridoxine (Vitamin B6), 1 mg for each mg INH, if unknown amount ingested, give 5 mg IV; HCO3, for alkaline diuresis; consider dialysis

Water Hemlock *(Cicuta douglasii)* → intractable seizures occur 1hr after ingestion; ↑ mortality; Treatment = hemodialysis

Jimson Weed *(datura stramonium)*

Anti-epileptics that may Cause Seizures:
Lamictal (lamotrigine), Phenytoin, Fosphenytoin, carbamazepine (Tegretol)

CAUSES OF SEIZURES – WITHDRAWAL
Mnemonic - (BEGS)

B	**B**aclofen
E	**E**TOH
G	**G**amma-hydroxybutyrate (GHB)
S	**S**edative Hypnotics

TOXICOLOGY

NON-CARDIOGENIC PULMONARY EDEMA
Mnemonic - **(MOP CD)2 ASAP**

M	**M**ethamphetamine smoking meth	**M**ountain sickness (**HAPE**=**h**igh **a**ltitude **p**ulmonary **e**dema)
O	**O**piates (heroin, methadone)	**O**verwhelming Sepsis
P	**P**henobarbital	**P**ancreatitis

C	**C**arbon Monoxide / Chlorine	**C**VA (neurogenic pulmonary edema)
D	**D**rugs	**D**rowning

A	**A**SA (Salicylates - more common in chronic poisoning and adults more > common than peds)
S	**S**ympathomimetics (smoking cocaine or methamphetamines)
A	**A**PAP
P	**P**hosgene

ANTI-cholinergic toxicity = *cardiogenic* pulmonary edema secondary to depressed myocardial contraction

Choking Agents
Phosgene and Chlorine = non-cardiogenic-pulmonary edema
Delayed symptoms with phosgene; *immediate* symptoms are noted with chlorine

LITHIUM TOXICITY
[10, 6th ed., pgs 1048-1051; 5th ed., pg 2172]

- Hand tremor 65%
- N / V/ D, anorexia, abdominal cramping → common early
- Polyuria → nephrogenic diabetes insipidus (DI) → inhibits arginine vasopressin
- (Antidiuretic hormone, ADH); more common in chronic toxicity
- Antithyroid effect → blocks thyroid hormone release → myxedema coma
- Parathyroid hyperplasia or adenoma → mild ↑Ca 2+
- EKG = flat T waves, ↓ST, ↓HR, U waves, ↑QTc (inhibits Na/K pump → ↓K+ in cell), ventricular arrhythmias, sinus arrest, asystole
- CNS = lethargy, confusion, spasticity, cogwheel rigidity, ↑DTRs, coma and seizures
- Acute overdose more GI and less CNS toxicity
- Normal range 0.6 to 1.2 mEq/L (lithium is slowly eliminated from cells, half-life = 24+12 hours, therefore patients may be toxic with levels in therapeutic range)
- Serum levels especially do not predict CNS levels (brain may → 2 to 3x higher concentrations)

LITHIUM TOXICITY TREATMENT

Mild/moderate toxicity
- Aggressive hydration with **0.9% NS** (if dehydrated kidney reabsorbs lithium preferentially)
- **Kayexalate** 15gm po (risk for hypokalemia)
- **Charcoal** is *ineffective* → consider charcoal in polydrug toxicity
- **Whole-bowel irrigation** → helpful, especially if sustained-release lithium products

TOXICOLOGY

Severe toxicity → **Hemodialysis**
- Coma, seizures
- Level > 3.5 or 4.0 mEq/L in acute overdose
- Patients with minimal change in their levels after 6 hours of hydration
- Patients with sustained levels > 1.0 mEq/L after 36 hours
- Seizures = BZP, Phenobarbital; Dilantin ↓ renal excretion of lithium and is ineffective
- If patient is asymptomatic after 6 hours of treatment of acute, non-sustained release, ingestions → psych consult

Interesting trivia provided by Dr. Bryan Bluhm → when the soft drink 7-UP was introduced in 1929 it originally contained lithium; fyi the molecular weight of lithium = 7

5 STAGES OF IRON TOXICITY
[12, 6th ed., pg 1122-1123]

Stage 1 (Gastrointestinal) Within 6 hours of ingestion
- N /V/ D abdominal pain; GI bleeding common

Stage 2 (Latent) usually occurs 6-12 hours after ingestion and may lasts 24 hours
- Resolution of GI symptoms
- This deceptive phase as the patient appears to improve and recover
- Metabolic abnormalities during this phase may include ↓BP, metabolic acidosis, and coagulopathy

Stage 3 (metabolic/cardiovascular – systemic toxicity) starts as early as 6-8 hours, lasts up to 2 days
- Most patients die during this phase
- Acute Renal Failure
- Recurrent GI bleeding
- Cardiomyopathy with CV collapse
- Coagulopathy – worsens bleeding; leukocytosis
- CNS symptoms = lethargy, encephalopathy and coma
- Anion-Gap metabolic acidosis (intracellular iron disrupts cellular metabolism)
- Combination of fluid and blood loss, with additional third-spacing → hypovolemia or shock

Stage 4 (hepatic) 2 to 5 days after exposure
- ↑ LFTs, BILI and ↑coagulopathy (hepatic injury → ↓factor production / hepatic failure)
- Hepatic injury → ↓ Glucose (hypoglycemia)
- ↑ Ammonia → encephalopathy, coma

Stage 5 (delayed) Usually → 4 to 6 weeks after a severe poisoning
- Gastric outlet obstruction or proximal bowel scarring / obstruction; rare

Moderate Toxicity = 20 to 60 mg/kg of elemental iron
Severe Toxicity = > 60 mg/kg of elemental iron

IRON TOXICITY TREATMENT
- Fluid, vitamin K, FFP, blood, antiemetics
- Consider gastric lavage if presentation within 1st hour
- Whole-bowel irrigation
- EGD to remove pills
- Deferoxamine 90 mg/kg IM (up to 1 gm in PEDS), q 4 to 6 hours as clinically indicated; binds directly to free iron; Change in urine color (vin rosé urine) → should not be the sole factor in deciding toxicity.

TOXICOLOGY

TRICYCLIC ANTIDEPRESSANT (TCA) PATHOPHYSIOLOGY
[10, 5th ed., pg. 2092-93]

1) **Inhibition of amine uptake** → ↑ serotonin → serotonin syndrome and ↑ norepinephrine, dopamine → early sympathomimetic effects (↑ HR, early mild HTN, followed by hypotension, arrhythmias)

2) **Anticholinergic effects** → only muscarinic, NOT nicotinic → central anticholinergic symptoms delirium, hallucinations, seizures, sedation and coma peripheral symptoms please see anticholinergic toxicity

3) **Inhibition of adrenergic post-synaptic receptors (α1 and α2)** → ↓ BP and reflex ↑ HR. Inhibition of ocular α adrenergic receptors → **miosis** which frequently offsets anticholinergic-induced **mydriasis**

4) **Na+ channel blockade** → Bradycardia, QRS prolongation, RAD, hypotension & seizures

    Na+ channel blockade → negative chronotropic → ↓ **HR** (the ↑ HR from anticholinergic activity partially offsets the ↓ HR; if patients has ↓ HR and **wide QRS** = ↑ toxicity/Na channel blockade)

    Na+ channel blockade → ↓ Na+ influx and delayed depolarization → ↑ **QRS & PR** on EKG

    ↓ rapid Na+ influx → ↓ release of intracellular Ca++ → ↓ myocardial contractility → ↓ **BP**

    Na+ channel blockade → (other: VT/VF, heart blocks, **seizures**)

5) **K+ channel antagonist** → ↓ K+ efflux during repolarization → ↑ *QT* → torsades

6) **GABA antagonism** → **seizures**

CARBON MONOXIDE (CO) POISONING – PATHOPHYSIOLOGY

- The affinity of CO for hemoglobin is 250x > than the affinity of O2 for hemoglobin → ↓ of the O2 carrying capacity in the blood → tissue hypoxia

- Shifting of the oxygen-hemoglobin dissociation curve → **LEFT** → ↓ O2 available at cellular level → ↓ tissue hypoxia

- Binds to cytochromes A, and P450 → inhibit cellular (mitochondrial) respiration → lactic acid

- Binds to myoglobin → ↓ the O2 available to myocardium → ischemia / arrhythmias

- Brain lipid peroxidation → neuronal damage

- See mnemonics, atrial fibrillation, seizures and non-cardiogenic pulmonary edema

- Classic *cherry red* skin is rarely seen

- Visual disturbances are frequent and *correlate with the duration of exposure.* Flame-shaped retinal hemorrhages, bright red retinal veins, papilledema, blindness is uncommon.

- Rhabdomyolysis → renal failure

TOXICOLOGY

- Check EKG, cardiac enzymes, ABG, lactate and COHB level

- N/V/D, hepatic necrosis, hematochezia, melena, non-pancreatic ↑amylase, DI, ↑glucose, ↓Ca, ↑BUN/Cr, DIC, ↑WBC

- CO is removed almost exclusively via the pulmonary circulation through competitive binding of hemoglobin by O2

- Serum elimination half-life of carboxyhemoglobin (COHb)
  Remember: ↓ 1/3 → 1/3 → 1/3
  Breathing room air = 180 minutes → 100% O2 = 60 min → 20 min with HBO [12, 6th ed., pg. 1240]
  Breathing room air = 300 minutes → 100% O2 = 90 min → 30 min with HBO [Up to Date]

HYPERBARIC OXYGEN (HBO) DEFINITE INDICATIONS
[12, 6th ed., pg. 1240]

- ↓BP
- Coma
- Seizure
- Myocardial ischemia
- ↑ Prolonged exposure
- LOC or near syncope
- Pregnancy with COHB > 15%; fetal distress
- MS changes and /or abnormal neuro exam

ACETAMINOPHEN (N-ACETYL-PARA-AMINOPHENOL, OR APAP) OVERDOSE PEARLS
[12, 6th ed., pg. 1088-1093]

- 90% of APAP is metabolized in the liver to **sulfate** and **glucuronide** conjugates.
  Conjugated metabolites lack biologic activity and are not hepatotoxic → excreted in urine

- 5% remaining APAP is → excreted unchanged in the urine and

- 5% remaining is metabolized via the hepatic **cytochrome P450 pathway** to a hepatotoxic metabolite → N-acetyl-p-benzoquinone-imine (NAPQI)

- The amount of NAPQI formed during the metabolism of APAP at therapeutic doses can be detoxified by conjugation with hepatic glutathione → nontoxic NAPQI conjugates → excreted in urine

- At *toxic doses,* acute ingestion = **150 mg/kg peds, 7.5 grams adults** → *sulfate* and *glucuronide* conjugation become saturated → ↑APAP is shunted into the cytochrome P450 pathway for further metabolism → ↑ NAPQI → **glutathione** stores are depleted → hepatic necrosis

- Drugs that **stimulate/induce the P450 pathway** may enhance APAP toxicity: (chronic use of: antihistamines, phenytoin, barbiturates, carbamazepine and chronic ETOH abuse)

- Drugs that **inhibit the P450 pathway** protect against APAP toxicity: cimetidine (Tagamet)

- *Acute* ingestion of ETOH protects against APAP toxicity by competitive inhibition via the cytochrome P450 pathway → ↓ amount of NAPQI produced

TOXICOLOGY

- **PEARL:** children are less susceptible to hepatoxicity than adults (increased rate of sulfation and increased relative size of liver affords hepato-protective effects to pediatric patients)

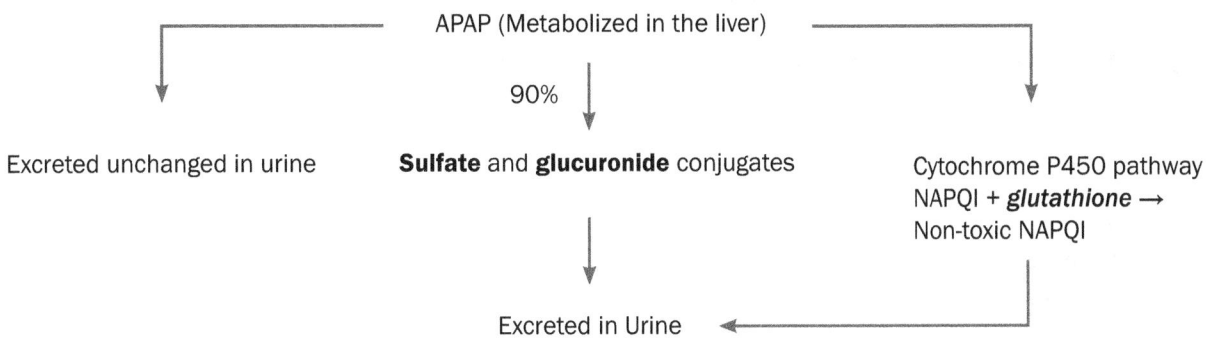

At **toxic doses** sulfate and glucuronide pathways become saturated → ↑APAP metabolized by cytochrome P450 enzymes. Once glutathione stores are depleted → ↑ NAPQI → hepatic injury

ACETAMINOPHEN (N-ACETYL-PARA-AMINOPHENOL, OR APAP) OVERDOSE PEARLS
[12, 6th ed., pgs. 1088-1093]

FOUR STAGES OF ACETAMINOPHEN POISONING

Stage	Time Following Ingestion	Characteristics
I	First day	Anorexia, N/V/malaise, lethargy, pallor, diaphoresis
II	1 to 3 days	Abdominal pain, liver tenderness, elevated LFT's, lipase, oliguria (ATN)
III	3 to 4 days	Peak LFT's and ↑ PT/INR, jaundice, confusion (hepatic encephalopathy), ↑ammonia, lactic acidosis Hypoglycemia
IV	4 days to 2 weeks	Resolution of hepatoxicity or progressive failure

- The serum APAP concentration should be measured at 4 and 24 hours after a single large overdose of an immediate-release preparation. The level should be evaluated according to the **Rumack-Matthew nomogram** to determine the risk of hepatotoxicity and need for therapy
- Other labs: serum tox, urine tox, pregnancy, UA, LFTs, lipase, coags, CBC, electrolytes, glucose, BUN/Cr; severe overdoses – ammonia level
- ETOH and chronic APAP poisoning hepatitis AST (GOT)/ALT (GPT) ratio > 2
- *Acute* APAP poisoning hepatitis AST/ALT ratio < 2
- An early marker for subclinical hepatic injury following APAP overdose is serum alpha glutathione S-transferase (a-GST), which is both released into and cleared from the circulation more rapidly than AST
- See mnemonic, "Drugs that cause pancreatitis"
- Elevated PT / INR indicate impaired synthetic liver function
- Non-cardiogenic pulmonary edema may develop in severe APAP poisoning

TOXICOLOGY

- The proximate cause of death in APAP-induced hepatic failure is usually cerebral edema
- APAP crosses the placenta, which places the fetus at potential risk of hepatotoxicity
- NAPQI does not cross the placenta
- The fetus metabolizes APAP → hepatotoxicity

ACETAMINOPHEN OVERDOSE PROGNOSTIC INDICATORS
[Mahadevan et al, 2006]

1. Delayed presentation to the emergency department
2. Delay in treatment
3. Prothrombin time > 100 seconds
4. Serum creatinine > 200 mcmol/L (2.26 mg/dL)
5. Hypoglycemia
6. Metabolic acidosis
7. Hepatic encephalopathy grade III or higher

Acetaminophen Antidote = Acetylcysteine; other names = N-acetyl-L-cysteine or N-acetylcysteine (NAC)
Trade names = Mucomyst (oral) and Acetadote (IV)

- If given within 8 hours NAC almost 100% effective in preventing hepatotoxicity
- Most toxicologists believe mucomyst prevents APAP hepatotoxicity by restoring hepatic glutathione stores; N-acetylcysteine also increases oxygen extraction by the liver, prevents WBC migration to the liver and acts as a free radical scavenger ---→ may be mechanism for improved hemodynamic and decrease in cerebral edema

Mucomyst (acetylcysteine)
1. **140 mg/kg po** loading dose
2. Next, give 70 mg/kg every 4 hours x 17 doses

Acetadote (acetylcysteine)
1. 150 mg/kg **IV** in 200cc D5W over 60 minutes
2. 50 mg/kg IV in 500cc D5W over 4 hours
3. 100 mg/kg IV in 1,000cc D5W over 16 hours

Activated charcoal
- Binds NAC, however the reduction in NAC absorption is insignificant
  Dose of charcoal = 1 gm/kg (max dose 50 mg)

Hemodialysis
- Because of its low volume of distribution and minimal protein binding, acetaminophen can be removed via hemodialysis
- However, considering N-acetylcysteine is so effective in the management of acetaminophen toxicity, the role for dialysis is *minimal*

Gastric lavage
- No proven efficacy in isolated acetaminophen overdose

TOXICOLOGY

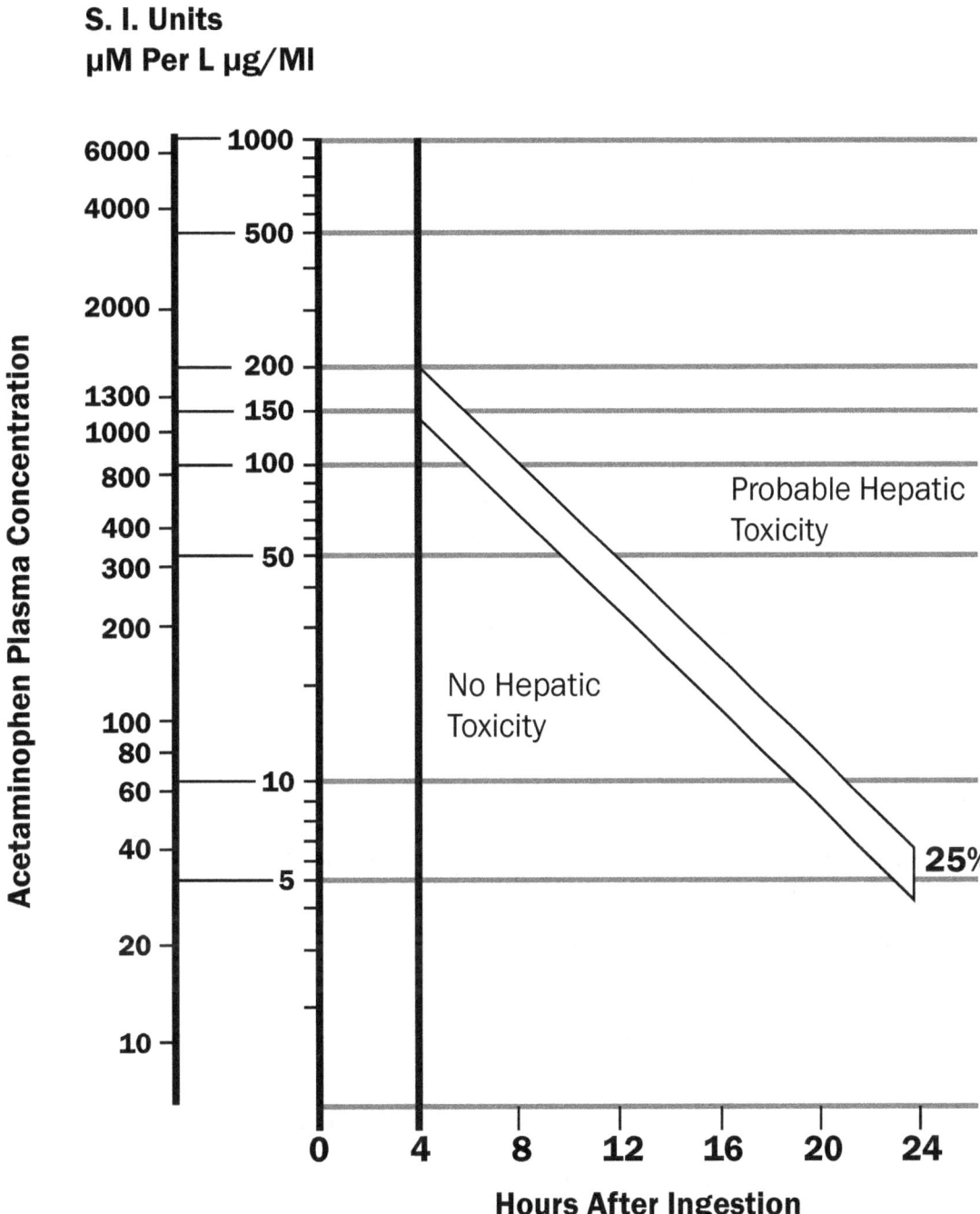

Rumack-Matthew nomogram for the single acute acetaminophen poisoning. Semilogarithmic plot of plasma acetaminophen levels versus time. Caution s for use of this chart: (1) The time coordinates refer to time of ingestion. (2) Serum levels drawn before 4 hours may not represent peak levels. (3) The graph should be used only in relation to a single acute ingestion. (4) The lower solid line 25% below the standard nomogram is included to allow for possible errors in acetaminophen plasma assays and estimated time from ingestion of an overdose. (Adapted from Rumack BH, Matthew H: Pediatrics 55:871-876, 1975.)

TOXICOLOGY

ACETAMINOPHEN OVERDOSE PEARLS
(N-acetyl-para-aminophenol, or APAP)

The 150 Rule
- Toxic dose is 150 mg/kg
- Give NAC if level is >150 mcg/mL four hours post-ingestion
- Initial loading dose of NAC is 150 mg/kg IV (140mg/kg PO)

SALICYLATE OVERDOSE
[12, 6th ed. pgs., 1085-1088]

- Acute ingestion of **150 to 300 mg/kg** produce mild/moderate toxicity = N/V/diaphoresis/↑RR/tinnitus (often first symptom reported)
- **300 – 500 mg/kg** = severe toxicity
- **> 500 mg/kg** = potentially lethal

SALICYLATE OVERDOSE PATHOPHYSIOLOGY

1) Krebs cycle inhibition → ↑ lactate → ↑ *metabolic acidosis*
2) Uncouples oxidative phosphorylation
   - → ↓ ATP production → release of energy in form of ↑ heat (fever)
   - → ↓ ATP production → *cerebral edema*
   - → ↓ ATP production → acidosis → *cardiac* depression / hypotension / VT/VF
   - → ↑ anaerobic glycolysis → *hypoglycemia* → ↑ lipid metabolism → ↑ ketones → ↑ *metabolic acidosis*
3) Stimulates respiratory centers in brain stem → Kussmaul breathing → *respiratory alkalosis*
4) ↑ pulmonary capillary permeability→ noncardiogenic pulmonary edema /**ARDS**
   a. More common in adults > peds
   b. More common in chronic > acute ASA poisoning
5) ↓clotting factor VII synthesis, → prolonged ↑ PT /INR

SALICYLATE OVERDOSE DONE NOMOGRAM
[12, 6th ed. pgs., 1085-1088]

- Predict the degree of toxicity after an acute single ingestion (cannot use if ASA ingestion within the last 24 hours or ingestion occurred over several hours or chronic salicylate poisoning or ingestion of enteric coated ASA tablets)
- Serum level must be drawn at least **6 hours** after ingestion
- [12, 6th ed. pg 1087] = Done nomogram has limitations, deceptive utility and *not* recommended

SALICYLATE OVERDOSE TREATMENT

- Charcoal
- Oxygen; avoid intubation. Fluid resuscitation unless pulmonary edema
- Dextrose drip if AMS, regardless of serum glucose
- Bicarb and K+ are needed to produce alkaline urine
- Why K+? → When Na+ is reabsorbed, the kidney preferentially secretes H+ ions into the tubular lumen rather than K+, you need K+ ions to compete with H+ ions → produce more alkaline urine

TOXICOLOGY

- Monitor ABG (keep arterial pH > 7.4), electrolytes and urine pH every 2 hours
- Bicarb 1 mEg/kg IV boluses until arterial pH > 7.4
- Continuous IV infusion of 1 L 5% dextrose in water, which is added 50 to 100 mmolNaHCO3 and 40 mmol KCL, started at 2x maintenance rate
- ASA is eliminated by renal excretion
- Ionized ASA cannot be reabsorbed and → excreted
- Alkaline urine favors the formation of ionized salicylate → ↑ excretion
- Hemodialysis

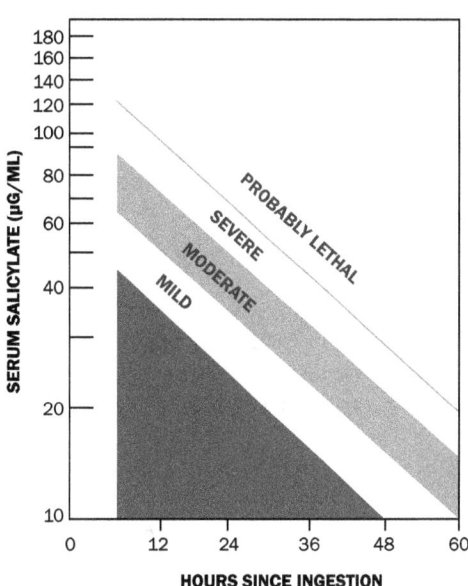

Pediatric Serum Salicylate Level and Severity of Intoxication Single Does Acute Ingestion Nomogram

METHEMOGLOBINEMIA

- A form of hemoglobin wherein the ferrous (Fe^{2+}) has been oxidized → ferric (Fe^{3+})
- The oxidized hemoglobin is incapable of carrying oxygen
- Causes: benzocaine, lidocaine, prilocaine, dapsone, sulfonamides, nitrofurantoin phenazopyridine (Pyridium), antimalarials (primaquine, chloroquine), nitrites and nitrates, acetaminophen, acetanilid, phenacetin, celecoxib
- Cyanosis is usually the first presenting symptom
- Blood sample = chocolate in color → turns red on exposure to air
- Treatment: Methylene Blue 1-2mg/kg, one 10 ml 10% solution (100mg) is initial adult dose[12, 6th ed., pg 1017]; treat if symptomatic or levels above 30%

Methemoglobin (MetHb) and carboxyhemoglobin (COHb) → ABG = normal PaO2 and calculated oxygen saturation, because the dissolved oxygen is unaffected, hence normal PaO2

The calculated oxygen saturation is based on the PaO2, therefore, will also be normal

You must order MEASURED oxygen saturation (measures the % hemoglobin bound to oxygen) – abnormally low ↓

TOXICOLOGY

MUSHROOM POISONING
[12, 6th ed., pgs 1242-1246]

- Mushrooms symptom onset
    - If onset < 6 hours after ingestion clinical course benign
    - \> 6 hours possible hepatotoxic, nephrotoxic, and erythromelalgic (hemolytic anemia) syndromes

- *Amanita* species (Cyclopeptides) = nearly **all** mushroom fatalities; 3 phases of illness:
  1) GI → 2) quiescent → 3) hepatic failure → hepatorenal syndrome
  Onset of symptoms delayed 6 to 24 hours; does not cross placental barrier
  No specific antidote, however treatment options include:
    - Activated charcoal
    - N-acetylcysteine (NAC)
    - High dose PCN G = blocks uptake of amatoxin in liver
    - Silymarin (silibinin) /Milk Thistle oral = hepatoprotective (available in Europe) – occupies receptor sites
    - Cimetidine, HBO
    - Liver transplant

- *Gyromita esculenta* → CNS and hepatotoxic; Methemoglobinemia; onset < 6 to 24 hours; heat labile; more toxic/fatal then Amanita however fewer fatalities because poisoning much less common
  Neuro symptoms treated with High Dose Pyridoxine (B6) – no effect on liver

- *Psilocybe* → structurally similar to LSD → psychedelic effects; onset < 30 min

- *Cortinarius orellanus* → norleucine toxin = nephrotoxic → delayed onset renal failure

- *Inocybe and Clitocybe* → muscarine → onset < 30 min SLUDGE BAM; tx = atropine

- *Amanita pantherina* (panther mushroom) → anticholinergic symptoms treatment of severe cases = physostigmine

- *Coprinus* → inky cap" or "shaggy mane" = disulfiram (Antabuse) reaction with ETOH onset 2 to 72 hours and < 30 min after ETOH (HA, flushing, SOB, ↑RR, ↑HR) [10, 5th ed., pg. 2203]
  Treatment = beta-blockers for SVT, Norepinephrine for refractory ↓ BP

- GI toxins with onset of symptoms < 2 hours = most commonly ingested mushroom
  N/V/D (occasionally bloody) / abdominal pain
  Chlorophyllum molybdates (green gill) – most common
  Omphalotus illudens (jack-o'-lantern)
  Boletus piperatus (pepper bolete)
  Agaricus arvensis (horse mushroom)

Hypoglycemia is one of the most common causes of death in mushroom toxicity

TOXICOLOGY

PEARLS - COCAINE, OPIATE, BARBITURATE, PCP, GHB, NMS, SYMPATHOMIMETICS

- **Cocaine** acts as a type IA sodium channel blocker → prolongs the QT

- **Cocaine** overdose: beta-blockers are contraindicated → unopposed a-adrenergic receptor stimulation → will worsen cocaine-induced coronary and peripheral vasoconstriction

- **Cocaine** use cause 90% of strokes in young adults (3rd and 4th decades of life), and is the most common cause of drug-associated stroke

- **Opiate**-induced pulmonary edema: treatment with diuretics is not effective. Use of naloxone and supportive treatment is all that usually is needed, and typically clears rapidly in 24-36 hours

- Miosis is a well-known side effect of opiate use however **meperidine** causes mydriasis instead

- Naloxone (Narcan) = 0.4 mg IV initial test dose should be given (avoid violent withdrawal symptoms). If no response is observed, give 2 mg doses q three minutes up to 10 mg total IV

- Clinical effects of **opiate** reversal with naloxone = 30-60 minutes → monitor for return of s/s of OD

- Cutaneous bullae occur in 4-6% of patients with **barbiturate** coma and in 50% of patients who die from barbiturate overdose

- Haloperidol (Haldol) use in **PCP** overdose may trigger a syndrome similar to neuroleptic malignant syndrome; consider benzodiazepines for sedation to avoid this potential complication

- Bruxism, or jaw clenching, is seen in nearly 100% of **MDMA** users. They often resort to use of pacifiers or lollipops to relieve jaw tension

- Hallmark of **GHB** overdose = rapid and profound CNS depression; deep levels of anesthesia last only 1-4 hours and spontaneously resolve with only supportive treatment

- **Sympathomimetics** present with similar symptoms as **anti-cholinergic** excess…. however, sympathomimetics → sweating and + bowel sounds

- Both **NMS** and **Serotonin Syndrome** = autonomic instability, altered MS and muscle rigidity

- NMS medical treatment options – controversial, include:
    - Dantrolene, a direct-acting skeletal muscle relaxant; 1 to 2.5mg/kg IV in adults, can be repeated to a maximum dose of 10 mg/kg/day
    - Bromocriptine, a dopamine agonist, restores lost dopaminergic tone. 2.5 mg (through nasogastric tube) every six to eight hours are titrated up to a maximum dose of 40 mg/day [http://www.uptodate.com/contents/neuroleptic-malignant-syndrome#H26]

- Rigidity in NMS more severe ("lead pipe" rigidity) >>> vs Serotonin Syndrome

- Hyperreflexia and clonus can coexist in Serotonin Syndrome, which is NOT typical in NMS

TOXICOLOGY

PEARLS - TODDLER TOX KILLERS

- Even a small amount of following can result in death: B-blockers, especially Inderal (propranolol), CCBs, camphor, ethylene glycol (tx dialysis), lomotil (tx naloxone), Amanita phalloides, Methyl salicylate, sulfonylureas (D5W & octreotide), TCAs and theophylline [Dr. Rangan, ACEP News, Vol. 30, No. 2, Feb 2011]

- Liquid nicotine commonly used in e-cigarettes now recognized as "one pill killer" [Journal of Paediatrics and Child Health Volume 50, Issue 2, pages 164-165, February 2014]

SULFONYLUREA OD, WITHDRAWAL SYNDROMES, CLONIDINE, LEAD, HEMLOCKS, DIG

- Treatment option of refractory hypoglycemia after sulfonylurea OD = **Octreotide** → somatostatin analog → inhibits insulin secretion from the pancreas that are a result of both the sulfonylurea and dextrose

- Acute ETOH **withdrawal** = anxiety, tremulousness, agitation, autonomic hyperactivity (↑HR, HTN, arrhythmias), hallucinations, and/or seizure

- Heroin withdrawal = yawning, N/V/D, abdominal pain, piloerection, restlessness, mydriasis and rhinorrhea

- Cocaine withdrawal = simulates depression

- **Clonidine** toxicity = ↓BP, ↓HR, ↓ RR, MS change, and miosis → closely mimics opioid toxicity

- Bluish lines on the gingival = **lead** lines from lead toxicity

- Lead concentrates in metaphyses of growing bones: distal femur, both ends of tibia or distal radius

- **Aldrich - Mees Lines** are horizontal lines of discoloration on the nails of the fingers and toes Poisonings = arsenic, thallium; or renal failure

- Water Hemlock (*Cicuta douglasii*) → intractable seizures occur 1 hour after ingestion; ↑ mortality

- Poison Hemlock (*Conium maculatum*) was reportedly used to execute Socrates (I had to add a Greek pearl)

- Root contains the greatest concentration of toxin in both species above

- **Digoxin toxicity** presentation = weakness, fatigue, nausea/vomiting/diarrhea, confusion, and a visual disturbance hallmarked by yellow/green halos around objects [12, 5th ed., pg. 1140]

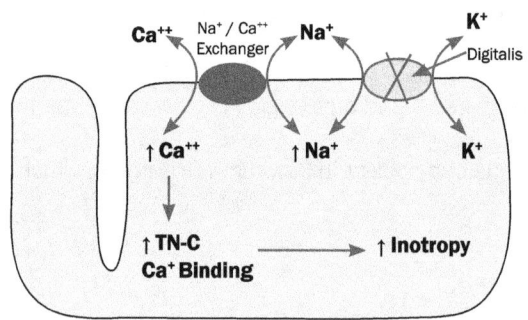

TOXICOLOGY

PEARLS - BETA-BLOCKER, CCBS, SEIZURES, ANESTHETICS, STRYCHNINE, ODORS

- Treatment options for **Beta-blocker and CCBs** Overdose
    - Glucagon 3 mg IV (max 10mg), then begin infusion 3 to 5 mg/hour [Ref 29, 2010 AHA pg. 47, 63]
    - Glucagon → enhances Ca entry and usage in cell → ↑ cardiac inotropy
    - High dose insulin 1.0 unit/kg bolus with Dextrose bolus (0.5g/kg IV); Insulin drip 0.5 unit/kg/hr with D25W
    - With CCB OD give Ca Gluconate or Ca Chloride (central line)
    - Vasopressors
    - Milrinone (Primacor) or Inamrinone (Inocor)
    - Intravenous lipid emulsion
    - Charcoal
    - Whole bowel irrigation for sustained release preparations [Dr. Rangan, ACEP News, Vol. 30, No. 2, Feb 2011]

- Beta-blocker cause seizures (see seizure mnemonic) while CCBs *rarely* cause seizures; tx BZPs

- Children can have significant life-threatening toxicity from minor accidental ingestions of B-blockers/CCBs

- Treatment options for local anesthetic toxicity = Intravenous lipid emulsion [Ref 29, 2010 AHA pg. 47, 63]

- Most common cause of drug-induce seizures = Bupropion (Wellbutrin, Zyban)

- **Isopropyl alcohol** → "pseudo renal failure" = early clue for diagnosis. ↑ Cr, normal BUN and ↑ acetone level

- Intoxications that improve with **Narcan**: heroin, clonidine, tramadol, captopril, ethanol and valproic acid [10, 5th ed, pgs. 1791-1793]

- **Strychnine** poisoning resembles = tetanus infection; Strychnine blocks glycine receptor (glycine is an inhibitory neurotransmitter)→absorbed rapidly→ acute generalized seizure like skeletal muscles contractions [10, 5th pg. 1792]

- The most commonly abused substance among adolescents presenting to the ED = alcohol

- In non-diabetic children ages 2 to 10, the most common drug induced cause of hypoglycemia = alcohol (Other causes, Reyes, sepsis, aspirin, adrenal insufficiency, hypothyroidism)

- **Toxic odors:**
    - Fruity – isopropanol, DKA
    - Pear like – chloral hydrate
    - Garlic – arsenic, organophosphates, DMSO, selenium, Mustard agent
    - Mustard – Mustard (blister) agent
    - Rotten Eggs – hydrogen sulfide, sulfur dioxide
    - Fresh hay – phosgene
    - Wintergreen mint – methylsalicylate
    - Moth balls – camphor, naphthalene
    - Onion - Mustard (blister) agent

TOXICOLOGY

NEUROLOGY

TRUE EMERGENT CAUSES OF SYNCOPE
Mnemonic - (CRAPS) [16]

C	Cardiac arrhythmia
R	Ruptured AAA or Ruptured Ectopic
A	Aortic stenosis and IHSS
P	PE
S	SAH

CAUSES OF SYNCOPE
(OH MY HEAD AND VESSELS ARE IN PAIN)

O	Orthostatic hypotension
H	Hypovolemia

MY	MI

H	Hypoxia (CO/Anemia)
E	sEizure
A	Aortic stenosis
D	Drugs

AND	Anemia

V	Vasovagal
E	Ectopic
S	SAH
S	Sensitivity (Hyper)-Carotid Sinus
E	Electrolyte abnormally → dysrhythmia
L	Low SVR (sepsis)
S	Subclavian steal

ARE	Anxiety → vasovagal

IN	IHSS

PAIN	Psychiatric; Pericardial tamponade

EVALUATION OF SYNCOPE

- The most Sn and Sp test in evaluation of patient with syncope = H&P
- Cause of syncope determined by thorough H&P in 50 to 85% of patients [10, 5th ed., 173]

NEUROLOGY

SAN FRANCISCO SYNCOPE RULE (SFSR)
Mnemonic - **(CHESS)**

C	History of **C**HF
H	**H**ematocrit < 30%
E	Abnormal **E**KG
S	**S**hortness of Breath
S	Triage **S**BP < 90

A patient with any of the above measures is considered at high risk for a serious outcome (death, MI, arrhythmia, PE, CVA, SAH, significant hemorrhage.

CAUSES OF HEADACHE
Mnemonic - **(MGM Studios Say They Can Create Happy Positive Movies to C)** [ANK]

M	**M**igraine / Tension / Cluster = Causes of "Primary" HAs [12, 6th ed., pg 1377]
G	*Causes of Secondary HAs* **G**laucoma / Iritis / Optic neuritis
M	**M**eningitis / Encephalitis / Cerebral Abscess

Studios	**S**ubdural / Epidural **S**AH / ICH
Say	**S**inusitis
They	**T**emporal arteritis
Can	**C**O poisoning
Create	**C**arotid or Vertebral a. dissection
Happy	**H**TN
Positive	**P**seudotumor cerebri **P**ost LP
Movies	**M**astoiditis
To	**T**umor (**C**NS masses)
C	**C**avernous sinus thrombosis

HEADACHE PEARLS

Most common recurrent head pain syndrome = **Tension HA** [10, 5th ed., pg 1460]
Women > men; 75% population affected

Post Dural Puncture HA (PDPH)
- Most common complication of LP 40% [10, 5th ed., pg 1464]
- Aggravated in the upright position and diminished in the supine position
- The amount of time supine does not affect incidence of HA

NEUROLOGY

Factors to ↓PDPH
    a. Smaller diameter needles → will cause less leakage
    b. Insert needle bevel parallel to long axis of spine → will minimize dura fiber damage
    c. Using atraumatic needles or pencil-point needles (Whitaker or Sprotte)

Treatment if bed rest and analgesics fail = epidural blood patch (autologous blood clot)

TEMPORAL ARTERITIS (TA) / GIANT CELL ARTERITIS (GCA)
3 of the following 5 items must be present

1. Age > 50 years
2. New-onset headache or localized head pain
3. Temporal artery tenderness to palpation or reduced pulsation
4. Erythrocyte sedimentation rate (ESR) greater than 50 mm/h
5. Abnormal arterial biopsy

Complications of GCA/TA = visual disturbances, diplopia and blindness.
- Diplopia is due to opthalmoplegia usually of CN III or VI
- Blindness is sometimes *preceded* by a visual field cut, amaurosis fugax, blurred vision or diplopia
- Blindness in GCA/TA is due to posterior ciliary artery occlusion → ↓ blood flow to optic nerve → ischemia → blindness (less common blindness is due to central retinal artery occlusion or ischemic retrobulbar optic neuritis)
- Definitive Diagnosis is made by temporal artery biopsy. Biopsy should be done within 1 week of the initiation of steroid therapy
- Few studies exist regarding dosing protocols for corticosteroids
- Start prednisone 60 mg/day po in the ED
- Improvement of systemic symptoms typically occurs within 72 hours of initiation of therapy

MENINGITIS PEARLS

LP may be performed without Head CT in some patient groups during evaluation for meningitis [Clin Infect Dis. 2004;39:1267-1284] [Rev Neurol Dis. 2006;3(2):57-60]

- Age less than 60
- Immunocompetent
- No history of CNS disease
- No recent seizure
- Alert and oriented
- No papilledema
- No focal neurologic deficits

Signs suspicious for space-occupying lesions = papilledema and focal neurologic deficits

The Infectious Disease Society of America (IDSA) states in its 2004 guidelines that the administration of antibiotics for suspected bacterial meningitis should be "emergent" but does not specify a time frame. [Clin Infect Dis. 2004;39:1267-1284]; time = ASAP

The decision to perform a head CT before lumbar puncture should not prevent the immediate administration of antibiotics

A paper published in 1989 found that the average time to antibiotic administration for bacterial meningitis is **three** hours [Ann Emerg Med. 1989;18(8):856-862]

NEUROLOGY

Although the yield of CSF cultures and CSF Gram stain may be diminished by antimicrobial therapy given prior to LP, pretreatment blood cultures and CSF findings (i.e., elevated WBC count, diminished glucose concentration, and elevated protein concentration) will likely provide evidence for or against the diagnosis of bacterial meningitis

(More meningitis pearls – see Pediatrics Section and Neurology Pearls)

STROKE SYNDROMES

Anterior Cerebral Artery [ICEP 2009 Board Review and 12, 6th ed., pg 1384-1385]

- Paralysis of opposite leg > worse than arm
- Sensory deficit paralleling paralysis
- AMS; confusion
- Bowel or bladder incontinence

Middle Cerebral Artery (Most common)

- Paralysis of opposite body, arm, face > worse than leg
- Sensory deficit paralleling paralysis
- Hemianopsia = Blindness in lateral half of visual field
- Agnosia = inability to recognize objects
- Aphasia (receptive, expressive or both) is present if dominant hemisphere involved

Right-handed and 80% left-handed patients are left hemisphere dominant
Non-dominant hemisphere involved = inattention, neglect, dysarthic (difficulty articulating words) but not aphasic

Posterior Cerebral Artery

- Hemianopsia
- AMS
- Cortical blindness
- CN III paralysis
- Most ischemic strokes will not be seen on CT for at least 6 hours
- 80% of **strokes** are ischemic. Patients with Afib are 10 to 20x more likely to develop stroke, and the majority of these are embolic events
- 30 day mortality after stoke = 20 – 25%; in-hospital mortality = 15% [10, 5th ed., pg 1433]
- Most common cause of new onset seizures in elderly = strokes
- Acute painless *vision loss* → anterior circulation stroke
  Common Carotid a. → Internal Carotid a. → first branch = ophthalmic artery →
  CN II (optic) and retina
- Internal Carotid artery terminates by branching → Anterior and Middle Cerebral arteries at the Circle of Willis

NEUROLOGY

STROKE SYNDROMES
Mnemonic - Brainstem Stroke (5 D's) [ANK]

D	**D**iplopia
D	**D**ysphagia
D	**D**ysarthria
D	**D**izziness – vertigo, nystagmus
D	**D**rop attacks (syncope)

Contralateral loss of pain and temperature
Bilateral spasticity
Facial numbness / paresthesias

- **Wallenberg Syndrome (Lateral Medullary Syndrome)** = *Ipsilateral* absence of facial pain and temperature, with *contralateral* loss of these senses over the body; ataxia; Horner syndrome; **d**ysphagia and **d**ysarthria (ipsilateral CN V, IX, X, XI involvement)

BASILAR ARTERY OCCLUSION
- Coma
- Severe quadriplegia
- *Locked in syndrome* (pontine lesion → complete muscle paralysis except for upward gaze)

CEREBELLAR INFARCTION
- N/V/HA/Neck pain
- Central Vertigo
- CN abnormalities often present
- Drop attacks
- 6 to 8 hours delay in cerebral edema → ↑ brainstem pressure → decrease LOC

CVA PEARLS
- Intracerebral lesions - gaze = TOWARDS the affected side
- Brainstem abnormalities - gaze = AWAY from the affected side
- Dysconjugate gaze (failure of the eyes to turn together in the same direction) [10, 5th ed., pg 141]
    - Vertical plane = Pontine or Cerebellar lesions
    - Horizontal plane = drowsiness, sedated states (ETOH intoxication)

STROKE MIMICS
Mnemonic - (MI HEMI) [http://www.jems.com/articles/print/volume-36/issue-3/patient-care/identifying-diseases-mimic-str.html with modification]

MI	**Mi**graine (hemiplegic migraine)

H	**H**ypo or **H**yperglycemia
E	**E**pilepsy (focal seizures)
M	**M**ultiple sclerosis
I	**I**nfections – CNS (encephalitis, meningitis or abscess)

NEUROLOGY

STROKE CAUSES IN YOUNG PATIENTS

Mnemonic - (MI HEMI 7 C's) [Step Up to USMLE Step 2, Van Kleunen JP. Lippincott, 2nd Edition 2008 from Dr Postel, with modifications]

1. **C**ocaine
2. **C**ancer (brain tumor)
3. **C**ardiogenic Emboli
4. **C**oagulation (sickle cell)
5. **C**NS infection (septic emboli)
6. **C**ongenital vascular lesion
7. **C**onsanguinity (genetic disease like Von Hippel-Lindau (VHL) syndrome, neurofibromatosis)

LACUNAR INFARCTS
[10, 5th ed., pg. 1434]

- Small vessel strokes (DM, HTN); 80-90% patients have HTN
- More common in African-American patients
- Most common sites: subcortical structures of cerebrum (BG, thalamus, internal capsule),
- & brainstem (pons)
- Most common lacunae syndromes = pure motor or **pure** sensory strokes or ataxic hemiparesis
- Subcortical, so **rarely** → cognitive deficits, aphasia, LOC, simultaneous motor-sensory findings or memory impairment

ALTEPASE (tPA) IN CVA

>18 y/o
Ischemic stroke
Time of symptom onset to drug administration < 4.5 hours [Stroke, 2009;40:2945-2948]
No contraindications to tPA

tPA / HEPARIN

tPA dose = 0.9 mg/kg, max 90mg
First 10% bolus over 1 min, remaining infused over next **60 min**
No not administer **heparin** or ASA during the first 24 hrs of fibrinolytic therapy [AHA, 2005 page 52]

CEREBRAL PERFUSION PRESSURE

CPP = MAP − ICP
Ideal CPP = > **70** mmHg
Normal intracranial pressure **(ICP) < 20 mmHg**

TIA

Focal symptoms, usually weakness or numbness, resolve within < 24 hours
Majority of TIAs last less than 30 minutes
(New proposed definition = A brief episode of neuro dysfunction caused by focal brain or retinal ischemia with clinical symptoms lasting < 1 hour without evidence of infarction)

NEUROLOGY

ABCD2
[Stroke. 2008 Nov;39(11):3096-8. Epub 2008 Aug 7]

A simple score to identify individuals at high early risk of stroke after transient ischemic attack

A	Age	1 point for age > 60 years
B	Blood pressure > 140/90 mmHg	1 point for hypertension at the acute evaluation
C	Clinical features	2 points for unilateral weakness 1 for speech disturbance without weakness
D	Symptom Duration	1 point for 10 to 59 minutes 2 points for > 60 minutes
D	Diabetes	1 point

Total scores ranged from 0 (lowest risk) to 7 (highest risk)

Stroke risk at 2 days, 7 days, and 90 days:
- Scores 0-3: low risk
- Scores 4-5: moderate risk
- Scores 6-7: high risk

Go to www.mdcalc.com/abcd2-score-for-tia → will calculate ADCD2 score for you and give you stroke risk at 2 days, 7 days, and 90 days

For example according to the validation study, an ABCD2 score of 4-5 points = Moderate Risk
2-Day Stroke Risk: 4.1%.
7-Day Stroke Risk: 5.9%.
90-Day Stroke Risk: 9.8%.

OCULOVESTIBULAR TESTING (COLD CALORICS) - DIRECTION OF FAST COMPONENT
Mnemonic - (COWS)

C	**C**old
O	**O**pposite
W	**W**arm
S	**S**ame

Exam of comatose patient (Test the integrity of the brainstem) →

Inject 50 cc ice water → TM, normal =
1) **Rapid** nystagmus *away* ("**Opposite**") from the irrigated ear
2) **Slow** compensatory nystagmus toward the irrigated side

- If the eye does not move in any direction, despite bilateral testing the brainstem is structurally or physiologically functionless
- Absence of oculovestibular reflexes = severe hypo-thermia, drug overdose and brainstem herniation. [12, 4th ed., pg. 228]
- If warm water is used = opposite will occur. The fast component of nystagmus = *toward* ("**Same**") irrigated side, and the slow component will be away from the irrigated side.

NEUROLOGY

OCULOCEPHALIC RESPONSE (DOLL'S EYES MANEUVER)

Tests integrity of **pontine gaze centers;** clear c-spine first; + if involuntary movement of the eyes upward/downward on passive flexion/extension of patient's head

GLASGOW COMA SCORE

Eye Opening (4)
4 = Spontaneous
3 = To voice
2 = To pain
1 = None

Verbal Response (5)
5 = Alert and oriented
4 = Disoriented conversation
3 = Speaking, but not coherent
2 = Moans or unintelligible words
1 = None

Motor Response (6)
6 = Follows commands
5 = Localizes pain
4 = Movement or withdrawal to pain
3 = Decorticate flexion
2 = Decerebrate extension
1 = None

HORNER SYNDROME
Mnemonic - (SPAM)

Sympathetic nerve impulses are disrupted and the pupil constricts due to more parasympathetic than sympathetic stimulation [12, 6th ed., pg. 1463]

S	**S**unken eyeball (enophthalmos)
P	**P**tosis
A	**A**nhidrosis
M	**M**iosis

Causes = internal carotid artery dissection, tumors, CVA, Pancoast tumor, herpes Zoster, trauma, Wallenberg Syndrome

ACUTE TRANSVERSE MYELITIS (ATM)

1. Back pain
2. Weakness of legs and arms
3. Transverse sensory impairment
4. Bowel and bladder dysfunction (incontinence or constipation/ retention)

- 1-4 new cases per million people per year
- All ages with bimodal peaks between the ages of 10-19 and 30-39 years of age
- 1/3 = no sequelae; 1/3 = moderate degree and 1/3 = severe degree of permanent disabilities
- Rapid progression = poor recovery
- Inflammation or demyelination of the spinal cord
- Several segments involved
- Bilateral findings
- Thoracic cord region 60-70%
- Causes: unknown, 30% follow viral infection [10, 5th ed. pg1501]
- Order spine MRI and CSF (elevated protein)
- Fever /back pain, ddx = spinal epidural abscess (may skip a level)
- Back pain with neuro symptoms ddx = malignancy or hematoma
- ↑ reflexes, sensory > weakness, clonus and Babinski ddx = MS (consider adding brain MRI)

NEUROLOGY

INCOMPLETE SPINAL CORD LESIONS

Central cord syndrome: usually in the elderly when the neck is subjected to hyperextension The ligamentum flavum buckles into the cord → contusion to the central portion of the cord → neurological deficits in *upper* extremity >>> lower extremities

Mnemonic - (MUD-E) [http://boringem.org/2015/11/16/a-boring-guide-to-spinal-cord-syndromes/]

M	**M**otor > Sensory
U	**U**pper extremity > Lower extremity
D	**D**istal > Proximal
E	**E**xtension injury

Anterior cord syndrome: flexion injuries → anterior cord compression → paralysis and pain-temperature loss distal to the lesion. **Preservation** of the Posterior Columns (fine touch, conscious proprioception and vibratory sense)

Brown-Sequard syndrome: hemisection of the spinal cord usually from penetrating GSW or knife wound, may be seen in lateral mass fractures of cervical spine →
- *ipsilateral* motor paralysis and loss of position-vibratory sensation
- *contralateral* sensory (pain and temperature) loss

SPINAL CORD TRACTS
Mnemonic - (SCALP)

	TRACT	FUNCTION	SITE OF CROSSOVER
S	**S**pinocerebellar tract	Muscle tone Unconscious proprioception	Ipsilateral
C	**C**orticospinal tract	Voluntary motor	Medulla
A	**A**nterior Spinothalamic	Crude touch	Spinal Cord
L	**L**ateral Spinothalamic	Pain Temperature	Spinal Cord
P	**P**osterior columns	Fine touch Conscious proprioception Vibratory sense	Medulla

- The maximum neurologic deficit after blunt spinal cord trauma is seen over many hours and is not seen immediately
- Factors that worsen spinal cord injury [10, 5th ed., pg 345]
  Hypoxia, hypoglycemia, ↓ BP, hyperthermia, mishandling by medical personnel

CAUDA EQUINA SYNDROME

- The spinal cord ends at L1-L2 vertebrae → The most distal of the spinal cord = conus medullaris, distal to this → collection of horsetail-like nerve roots = cauda equina (Latin for horse's tail) → Nerve root injury rather than a true spinal cord injury
- Most consistent finding = urinary retention [10, 5th ed., pg 1499]

NEUROLOGY

- Present with fecal or urinary incontinence (overflow), impotence, distal motor weakness and
- Sensory loss = saddle distribution over the perineum
- Common cause is ruptured disc, most common L4-L5; other causes = tumors, trauma, vascular
- Treatment = surgery; steroids controversial

COMPLETE SPINAL CORD SYNDROME ACUTE OR SUBACUTE

Total loss of sensory, autonomic and voluntary motor distal to spinal cord level of injury
DTRs present, may be ↑or ↓

Spinal Shock triad → ↓ BP, ↓HR, and peripheral vasodilation resulting from autonomic dysfunction and the interruption of Sympathetic Nervous System control

Spinal Shock → loss of bulbocavernous reflex (reflex contraction of anal sphincter in response to squeezing the glans penis or tugging on an indwelling Foley catheter)
- Associated head injury occurs in about 25% of patients with spinal cord injury
- Judicious fluids – avoid pulmonary edema; atropine for bradycardia; rarely dopamine

INSUFFICIENT EVIDENCE TO SUPPORT STEROIDS IN CORD INJURY
ATLS 8th Ed, 2008 [J Trauma 2008;64:1638-1650]

In case old school doc /institution requests / or test questions lag ... the steroid protocol →
Steroids: Treatment must be started within 8 hours of injury [12, 6th ed., pg. 1581]

1) Methylprednisolone 30 mg/kg bolus over 15 minutes
2) Followed by 45 minutes pause
3) Infusion of Methylprednisolone at 5.4 mg/kg/h is continued for 23 hours

Steroid complications: thromboembolism, sepsis, pneumonia, wound infection, delayed healing, GI bleed

CRANIAL NERVES
Mnemonic - (Oh, Oh, Oh, To Touch A Funky Vest Gives Very Amazing Happiness)

O	I.	**O**lfactory nerve
O	II.	**O**ptic nerve
O	III.	**O**culomotor nerve – EOMs, pupillary constriction via PNS, upper lid elevation (levator palpebrae)
T	IV.	**T**rochlear nerve (LR6SO4) → superior oblique mm → eye downward and laterally Patients compensate for CN IV compression by head tilt
T	V.	**T**rigeminal nerve – chew, face/mouth touch and pain
A	VI.	**A**bducens nerve – lateral rectus muscle → eye laterally
F	VII.	**F**acial nerve – face muscles, tears, saliva, taste
V	VIII.	**V**estibulocochlear nerve/Auditory nerve – hearing, equilibrium
G	IX.	**G**lossopharyngeal nerve – taste, senses carotid BP
V	X.	**V**agus nerve – senses aortic BP, slows HR, stimulates digestive organs, taste, Unilateral palatal elevation
A	XI.	**A**ccessory nerve/Spinal accessory nerve – trapezius, SCM, swallowing
H	XII.	**H**ypoglossal nerve – tongue motor

NEUROLOGY

CORD INJURY SENSORY LEVELS

Clavicle	C5
Thumb	C6
Index & middle finger	C7
Ring & small finger	C8
Nipples	T4
Umbilicus	T10
Medial thigh	L2-3
Knee	L4
Lateral calf	L5
Perineum	S2,3,4

NERVE INJURIES
Mnemonics - (DR CUMA) [ANK]

D	**D**rop wrist
R	**R**adial nerve
C	**C**law hand
U	**U**lnar nerve
M	**M**edian nerve
A	**A**pe hand

MEDIAN NERVE

1) Wrist flexion and the Flexor Digitorum Superficialis (FDS), finger flexion
2) *Median* nerve branches → to *anterior interosseus nerve* (controls deep finger flexors in the forearm - FDProfundus, Flexor Pollicis Longus, and Pronator Quadratus
3) → *sensory* branch provides sensation to most of palm plus a motor branch →
4) → *recurrent branch of the median nerve* → innervates thenar motor muscles of the thumb

Evaluation of Median Nerve
Test *anterior interosseus* nerve → make a circle or, "OK" sign, with thumb/index finger
Test *recurrent branch of median nerve* →ABDuction of the thumb
Sensation on radial side of palm – from the thumb to radial half of ring finger
Eponym "LOAF weakness" = **L**umbricals (flex MCP), thumb: **O**pposition, **AB**Duction, **F**lexion

ULNAR NERVE

Controls intrinsic muscles and provides sensation to the little finger and ulnar half of ring finger
Eponym for ulnar n palsy "Tardy Ulnar Palsy" – palsy can manifest years after injury.

NEUROLOGY

RADIAL NERVE

Wrist extension then branches → posterior interosseus n. → finger extension
Dorsal hand sensation. Test sensation over the dorsum of the thumb-index finger web space
Eponym for radial n palsy = "Saturday Night Palsy " and "Bridegroom Palsy"

LIFE THREATENING CAUSES OF ALTERED MENTAL STATUS
Mnemonics - (WHHHIMP) [ANK, 14]

W	**W**ernicke's encephalopathy (give thiamine)
H	**H**ypoglycemia
H	**H**TN encephalopathy
H	**H**ypoxia
I	**I**ntracerebral hemorrhage
M	**M**eningitis
P	**P**oisonings

CAUSES OF COMA
Mnemonic - (AEIOU SET TIPS) [ANK, 12, pg. 230 with modifications]
Note: AEIOU is the mnemonic used for "Dialysis Criteria"

A	**A**lcohol and other drugs – (opiates)
E	**E**ndocrine (↓↑ glucose, myxedema coma -↓ T3)
I	**I**ncreased BP or NH3 (Hypertensive or Hepatic Encephalopathy)
O	**O**xygen
U	**U**remia

S	**S**AH/Stroke
E	**E**lectrolytes (Na+, Ca++)
T	**T**rauma

T	**T**emperature (heat stroke)
I	**I**nfection
P	**P**sychiatric
S	**S**pace occupying lesions Shock Seizure → Post Ictal

NEUROLOGY

ALTERED LEVEL OF CONSCIOUSNESS TREATMENT OPTIONS
Mnemonics - (DONT) [28 Volume 31, Number 12]

D	**D**extrose - one AMP of D 50%
O	**O**xygen
N	**N**arcan (Naloxone) 2 mg IV
T	**T**hiamine 100 mg IV, give BEFORE glucose

CAUSES OF PERIPHERAL NEUROPATHY
Mnemonics - (HIT DANG SPARTEN) [ANK]

H	**H**ereditary
I	**I**nfectious (Diphtheria, Mono, syphilis, hepatitis, HIV)
T	**T**oxic (heavy metals: lead, arsenic, mercury, thallium)

D	**D**iabetes
A	**A**lcohol
N	**N**utritional Deficiencies: Thiamine (B1), Niacin (B3), Cobalamin (B12), Vit E
G	**G**uillain-Barre syndrome

S	**S**ystemic (SLE, PA, sarcoid, hypothyroid)
P	**P**orphyria
A	**A**myloid
R	**R**enal failure
T	**T**rauma
E	**E**maciation (see nutritional above)
N	**N**o known cause

NORMAL PRESSURE HYDROCEPHALUS - TRIAD
Mnemonic - (Wet - Wacky Wobbly) [ANK]

1) Wet   (Urinary incontinence)
2) Wacky   (Mental confusion)
3) Wobbly   (Ataxia)

HA and papilledema are **ABSENT**. Drop attacks may occur. Normal pressure hydrocephalus may follow SAH, meningitis or head trauma, although the cause is usually *unknown* [10, 4th ed. pg 2223]

PSEUDOTUMOR CEREBRI (IDIOPATHIC INTRACRANIAL HYPERTENSION)
Most frequently in young, overweight women between the ages of 20 and 45
Headache is the most common presenting complaint; papilledema on exam
Causes = pregnancy, medications (OCPs, steroids, vitamin A) [10, 5th ed., pg., 1518]

NEUROLOGY

TABES DORSALIS
Progressive demyelination of *Posterior Column* and *Dorsal Nerve Roots*. (Know how to distinguish Tabes Dorsalis from Normal Pressure Hydrocephalus)

1) Ataxia
2) Urinary incontinence and loss of sexual function
3) Leg pain (Lancinating - appearing suddenly, spreading rapidly, and disappearing) often is an early symptom

Other neurologic presentations: progressive loss of pain sensation, loss of peripheral reflexes, impairment of vibration and position senses

NEUROLOGY PEARLS

- Status epilepticus has an associated mortality of up to 20%

- Bell's Palsy = CN VII Palsy; lose ipsilateral forehead strength; steroids, antivirals [12, 6th ed. pg 1421]

- Most common CN involved in cephalic tetanus = CN VII (facial nerve)

- If you suspect SAH and head CT is normal → perform LP; note: 12 hours before you can see xanthochromia

- Possible complications weeks after **SAH**: seizure, cerebral artery vasospasm, hydrocephalus, rebleeding

- Basilar skull fracture = 1) Raccoon eye (periorbital ecchymosis) 2) Battle sign (mastoid ecchymosis) 3) hemotympanum 4) CSF rhinorrhea/otorrhea.

- Spinal shock = hypotension and bradycardia; Tx = Trendelenburg position and fluid

- ↑ICP → Cushing Reflex = ↑ BP and ↓ HR, with respiratory irregularities

- Most common presenting symptom of Multiple Sclerosis (MS) = optic neuritis

- Most common cause of neonatal (0-28 days) meningitis = *Group B Strep, E. Coli* and *Listeria* [12]

- Most common cause of meningitis 1st – 3rd month of life = GBS, *Listeria* and now include *Haemophilus, Strep pneumo* and *Neisseria meningitidis* [12]

- Most common cause of meningitis 3 months-18 years = *Strep pneumo, Haemophilus,* and *Neisseria* [12]

- Most common cause of **focal encephalitis** in patients with AIDS = *Toxoplasma gondii* (obligate intracellular parasite). Contrast CT → ring-enhancing lesions with surrounding areas of edema; seizures common and 80% focal neurologic deficits [10, 5th ed. pg 1849]
Treatment = Pyrimethamine + Folinic acid (prevents heme toxicity from pyrimethamine) + Sulfadiazine or Bactrim IV [Sanford, 2009, pg. 129]

- Most common opportunistic CNS **fungal** infection = *Cryptococcus neoformans*; ↑↑ ICP; seizures uncommon CSF cryptococcal antigen = 100 % Sn & Sp; treatment = Amphotercin IV + Flucytosine po [Sanford, 2009, pg. 103]

- HIV encephalopathy or AIDS Dementia Complex = 1/3 patients with HIV; early = impairment in recent memory, difficulty concentrating; later MS changes and seizures [10, 5th ed. pg 1849]

NEUROLOGY

NEUROLOGY PEARLS (CONT)

- *Listeria monocytogenes* risk > 60x ↑ in AIDS patients and 3/4 present = meningitis [Sanford, 2009, pg. 10]

- > 1st year of life, you can assess nuchal rigidity. Two signs of meningeal irritation = Kernig's and Brudzinski's

- Kernig's sign with patient lying supine, hip and knee flexed to about 90o, knee extension → in meningeal irritation → neck pain [12, 6th ed., pg. 741]

- Brudzinski's = with the patient supine, passive flexion of the neck → involuntary flexion in the hips, if there is meningeal irritation [12, 6th ed., pg. 741]

- DeCORticate posturing: hyperextension of legs with flexion of the arms; remember arms are in flexion with hands over the heart ("**cor**") in de "**cor**"ticate posturing – results from damage to the descending motor pathways above the central midbrain [9]

- Decerebrate posturing: hyperextension of both upper and lower extremities – refers to damage to the midbrain and upper pons; is a grave sign [9]

- Meniere's Disease = 1) Vertigo 2) Unilateral diminished hearing and 3) Tinnitus intermittent
Ddx acoustic neuroma = tinnitus is **constant** with acoustic neuroma; Meniere's = true vertigo vs acoustic neuroma patient describes imbalance and dysequilibrium

- **Peripheral Vertigo:** Nystagmus = unidirectional and fatiguing and suppressed with fixation. Usually have hearing loss. Caloric testing shows *abnormal* function on involved side. N/V/diaphoresis common

    Examples: Benign paroxysmal positional vertigo (BPPV), Meniere's disease, vestibular neuronitis, perilymph fistula, labyrinthitis [12, 6th ed, pgs. 1404-1408]

    BPPV, vestibular neuronitis → NO hearing loss

    Treatment options: 1) Sedative: diazepam 2) Antiemetic: hydroxyzine (Vistaril), promethazine (Phenergan), metoclopramide (Reglan) 3) Anticholinergic: Scopolamine patch 4) Anti-histamines: Antivert, Benadryl   5) Calcium antagonists: Nimodipine, Cinnarizine, Flurarazine [Table 231-4 in ref. 12, 6th ed, pg. 1404]

- **Central vertigo** (disorders affecting the cerebellum or brainstem): Nystagmus = multidirectional and Non-fatiguing and not suppressed with fixation. Caloric testing often normal

    Associated N/V/diaphoresis = **rare** and hearing loss is **unlikely**
    Examples = brainstem stroke, multiple sclerosis, acoustic neuroma, cerebellar stroke, Wallenberg Syndrome, vertebrobasilar insufficiency, vertebral artery dissection.

- The most common cause of delirium in the elderly = medications, 22-39% of cases. Delirium symptom onset acute vs Dementia progressive over months. Alterations in sleep-wake cycles are common
Delirium = various hallucinations (visual, auditory, olfactory, tactile, gustatory) vs
Functional Psychosis patient who only experience auditory hallucinations
EEG = abnormal; bilateral diffuse symmetric abnormalities; relative generalized slowing with or without superimposed fast activity [10, 5th ed., pg. 1471, 6th ed., 1645-64]

- One of the hallmarks of Acute Delirium = short-term memory impairment; remote memory preserved

- Post-traumatic seizures: more common PEDS > adults; if dura disrupted the incidence of seizures with neuro deficits ↑; incidence of seizures ↑ than general population

NEUROLOGY

- Dilantin loading dose = 18 to 20 mg/kg IV with infusion < 50 mg/min

- Treatment for Trigeminal Neuralgia = oral Carbamazepine (Tegretol)

- Most common ocular motor palsy = **CN VI** (abducens nerve) innervates the ipsilateral lateral rectus
  Causes = aneurysm, vascular disease (DM, HTN, atherosclerosis), trauma, neoplasm, MS, MG, meningitis, cavernous sinus mass, thyroid eye disease, ↑ ICP → downward displacement of the brainstem (30% of patients with pseudotumor cerebri have an isolated abducens nerve palsy)

- **CN III (Oculomotor) Palsy causes** = Infarction, hemorrhage, neoplasm, aneurysm, abscess, meningitis, vascular disease (DM, microvascular ischemia/atherosclerosis), cavernous sinus mass, thyroid eye disease

- Pinpoint pupils = Pontine hemorrhage; also, see TOX - (COPS) 2

- SLE CNS manifestations = seizures, CVA, psychosis, migraines and peripheral neuropathy

- Initial work-up of CSF shunt malfunction = CT head and shunt series (consider in any patient with shunt and decreased LOC)

- **Myasthenia gravis (MG)** = autoimmune disease - antibodies directed against the acetylcholine receptor at NMJ → diplopia and ptosis most common presenting symptom, also = dysphagia, dysarthria then muscle weakness
  Females > males; teens to early 30's; thymomas thought to ↑ acetylcholine receptor antibodies

  Tensilon test helps with diagnosis. Edrophonium (Tensilon) temporarily blocks acetylcholinesterase → prolongs muscle stimulation and temporarily improves strength (measure distance from the upper to the lower eyelid in the most severely affected eye before and after edrophonium). 1 –2 mg test dose

- Treatment of MG = **Anticholinesterase medications**
  - Neostigmine (Prostigmin) and Pyridostigmine (Mestinon)

- Lambert Eaton with repeated stimulation → increase strength, opposite of MG

- **Nerve Agents** (sarin, soman, tabun, GF, VX) are organophosphates which are potent inhibitors of acetylcholinesterase → SLUDGE BAM syndrome. Treatment = atropine and 2-PAM (pralidoxime)

- **Botulism** = neurotoxins → block the release of acetylcholine → descending symmetrical paralysis, ptosis, generalized weakness UE >LE, proximal muscles > distal, dizziness, dry mouth, diplopia, dilated/fixed pupils & blurred vision, dysphonia, dysarthria, dysphagia, and respiratory failure. ↓ DTR's

  Treatment = antitoxin → binds circulating toxin only; intubate if vital capacity <30%; saline enema to cleanse GI tract of residual toxin

- **Guillain-Barre Syndrome (GBS)** = ascending paralysis; in most cases GBS is caused by an autoimmune attack on myelinated motor nerves. Variable sensory findings; subjective sensory disturbances (numbness and tingling of lower extremities); urinary retention may occur -ddx = spinal cord lesion and cauda equina syndrome

- **GBS** – order vital capacity

- Hallmark of **GBS** = loss of deep tendon reflexes (↓ DTR's). CSF = high protein, normal glucose and cell count.
  Treatment = plasma exchange or intravenous immunoglobulin [12, 6th ed, pg. 1420]

- Most common CNS complication of pertussis = seizures 3%; other = encephalopathy and ICH

- **Tick Paralysis** = mimics GBS

HEMATOLOGY / ONCOLOGY

CAUSES OF MICROCYTIC HYPOCHROMIC ANEMIAS
Mnemonic - (TAILS) [ANK, provided by Dr. Matt Jordan]

T	**T**halassemia
A	**A**nemia of chronic disease
I	**I**ron deficiency
L	**L**ead Poisoning
S	**S**ideroblastic

CAUSES OF THROMBOCYTOSIS
Mnemonic - (Crazy Navy Men SHIP) [ANK]

Crazy	**C**irrhosis	**C**ML	**C**ollagen vascular Disease
Navy	**N**eoplasm (GI)		
Men	**M**yelofibrosis		

S	**S**plenectomy (post)			
H	**H**emophilia	**H**emorrhage (post)		
I	**I**nfection	**I**ron Deficiency	**I**BD	
P	**P**ancreatitis	**P**ostpartum	**P**ost trauma	**P**olycythemia vera

CAUSES OF EOSINOPHILIA
Mnemonic - (NAAACP)

N	**N**eoplasms
A	**A**llergy
A	**A**sthma
A	**A**ddison's
C	**C**ollagen vascular disease (CVD)*
P	**P**arasites

*CVDs = systemic lupus erythematosus (SLE), rheumatoid arthritis (RA), progressive systemic sclerosis (PSS) or scleroderma (SD), dermatomyositis (DM) and polymyositis (PM), ankylosing spondylitis (AS), Sjögren syndrome (SS), and mixed connective-tissue disease (MCTD)

Churg-Strauss syndrome (CSS), or allergic granulomatous angiitis = Eosinophilia; Wegener granulomatosis, another ANCA vasculitic syndrome similar to CSS = NO eosinophilia

HEMATOLOGY / ONCOLOGY

THROMBOTIC THROMBOCYTOPENIC PURPURA (TTP)
Mnemonic - (FAT RN)

F	**F**ever in 90%
A	**A**nemia – Microangiopathic hemolytic anemia (MAHA) – schistocytes on smear
T	**T**hrombocytopenia

R	**R**enal failure
N	**N**eurologic sequelae (HA, confusion, CN palsies, seizures and coma)

- Pentad only present in 40%
- ↑LDH, ↑IBIL, ↑Reticulocyte count, ↓haptoglobin, ↑Cr, normal DIC panel
- Abdominal pain
- Affect any age but majority 10 to 40 years
- Women 60% of cases
- 1/3 of patients who survive initial episode experience relapse within 10 years
- Survival rate = 80-90% with early diagnosis and treatment. 95% MR with no treatment
- Treatment = Intravenous (IV) plasma exchange, also called plasmapheresis

IDIOPATHIC THROMBOCYTOPENIC PURPURA (ITP)
Syn = primary immune thrombocytopenic purpura and autoimmune thrombocytopenic purpura [12, 6th ed., pg 1325]

- Patients having antibodies to platelet membrane glycoproteins
- Most often in children (usually 2 to 6 yrs); male = female; follows viral infection
- Usually **isolated** thrombocytopenia; M&M low; recovery may take weeks
- Treatment is supportive as the course is self limited with 90% spontaneous remission
- Platelet count < 20,000 to 30,000 µL require treatment
- Platelet count < 50,000 µL with bleeding *or* risk factors for bleeding require treatment
- Treatment = steroids; life threatening bleeding *high dose steroids + IV immunoglobulin*
- Add conjugated estrogen 25 mg IV x once if uterine bleeding
- Admit if ITP-related bleeding and consider admit if platelet count < 20,000 µL

HEMOLYTIC UREMIC SYNDROME (HUS)
Use TTP Mnemonic with Pearls below:

- Can occur at any age however majority of cases < 5 years old
- Low grade fever 5-20%; abdominal pain
- Renal manifestations more prominent than neurologic ones in HUS vs TTP
- 75% of cases post-infectious: *E. coli* O157:H7, *Shigella*, *Strep. pneumoniae*
- Majority = gastroenteritis up to 2 weeks before illness. *E. coli* O157:H7 [12, 6th ed., pg. 820]
- 15% children vs 5% adults infected with *E. coli* O157:H7 go on to develop HUS
- Mild HUS → steroid therapy may be beneficial
- Severe disease plasmapheresis = equivocal results [12, 6th ed., pg. 1347]
- Hemodialysis if renal failure (shorter duration = better chances of recovery from HUS)
- Infection with *E. coli* O157:H7 should *not* be treated w/ antimotility drugs → ↑risk HUS
- Antibiotics in treatment of *E. coli* O157:H7 = controversial [Sanford Guide, 2009, pg.16]

HEMATOLOGY / ONCOLOGY

METS TO BONE
Mnemonic - (**M**any **K**inds **o**f **T**umors **L**eaping **P**romptly **T**o **B**one) [ANK]

Many	**M**yelomas
Kinds	**Ki**dney
Of	**O**vary
Tumors	**T**esticular
Leaping	**L**ung **L**ymphoma
Promptly	**P**rostate
To	**T**hyroid
Bone	**B**reast

- Mets most commonly to pedicles on thoracic vertebrae
- Most common site of malignant spinal cord compression = thoracic vertebrae.
- Most common symptom of spinal cord compression = pain (WORSE recumbent position)
- Most common early finding of spinal cord compression = motor weakness
- Check CT, MRI or myelogram; Treatment options = steroids, radiation and surgery

HEME /ONC PEARLS

- Cancer with JVD, dyspnea →
    - **Superior vena cava syndrome** = facial plethora, dilated veins chest/arms, HA
    - Most common cause = Small (oat) cell lung CA > bronchogenic (squamous) cell lung CA > lymphoma; CXR = right sided mass; discuss with ONC – consider: radiation, steroids, lasix, endovascular shunts, and thrombolytics if thrombotic etiology
    - **Cardiac tamponade;** also↓BP
    - Massive **PE** → acute cor pulmonale; also ↓BP

- Cancer with back pain → **Spinal cord compression** treatment = Decadron 10mg IV, radiation

- Cancer with Paraneoplastic syndromes:
    - ↓Na+ → SIADH (euvolemic hyponatremia); small cell Lung CA
    - ↓Ca2+ → Calcitonin
    - ↑Ca2+ → Parathormone → "Stones (kidney), Abdominal groans (constipation, N/V, anorexia, pancreatitis), Psychic moans (mental status changes, seizures, coma, Bones (pain)"
    - ↑ACTH → Cushing syndrome (↑ cortisol), ↑Na+, ↓K+, (small cell Lung CA)
    - ↑ACTH → Gonadotropins / gynecomastia
    - ↑Serotonin → Carcinoid syndrome
    - Thymoma → ↑ acetylcholine receptor antibodies → Myasthenia gravis
    - Lambert-Eaton myasthenic syndrome (LEMS) → autoimmune attack on pre-synaptic motor nerve voltage-gated calcium channels (VGCC) → decrease in amount of acetylcholine in synapse → proximal muscle weakness, ↓DTR's & autonomic changes → (small cell Lung CA)

- FUO consider malignancy → lymphomas, acute leukemias, sarcomas, renal cell carcinomas, GI malingancies

- Neutropenic = neutrophil count < 500 /mm3

HEMATOLOGY / ONCOLOGY

- Emergent treatment of hyperviscosity syndrome secondary to symptomatic polycythemia = phlebotomy (not more than 500 ml of blood is removed and the volume is replaced with an equal volume of normal saline)
- Other Hyperviscosity Syndromes = Multiple Myeloma and Waldenstrom's Macroglobulinemia (sludge immunoglobulins) and Leukemia (sludge wbc's) tx = plasmapheresis or leukapheresis

- Tumor Lysis Syndrome = cell lysis → ↑ K+ (dysrhythmia), ↑Phos, ↑Uric Acid (renal failure) and subsequent ↓ Ca 2+ (muscle cramps/tetany); Treatment options:
Chemo pretreatment with fluid and allopurinol
Urine alkalinization controversial – improves uric acid diuresis, but may worsen ↓ Ca 2+ tetany
Dialysis = Treatment of Choice

HEME PEARLS

- Vitamin K dependent factors = 2, 7, 9, 10

- Extrinsic pathway = PT = Coumadin = 3, 7

- Cryoprecipitate = vWF, Fibrinogen, Factor 8 and Factor 13

- Prothrombin Complex Concentrate (PCC) (trade name Beriplex or Octaplex) = Vit K dependant coagulation factors II, VII, IX and X as well as protein C and S

- Desmopressin (DDAVP) = causes release of vWF
    - vWF → carries additional factor 8 in the plasma
    - vWF → is also required for normal platelet adhesion

- Fresh Frozen Plasma (FFP) = contains all the plasma clotting factors

- Hemophilia A = Factor 8 deficiency = most common cause of hemophilia in the US (↑ PTT)

- Hemophilia B = Christmas disease = Factor 9 deficiency (↑ PTT)

- Hemophilia A and B = X-linked recessive disorder – therefore a disease of men

- von Willebrand's disease = most common **inherited** bleeding disorder (autosomal dominant); ↑ bleeding time

- Most common cause of vaginal bleeding related to primary coagulation disorder = Von Willebrand's disease [12, 5th edn, pgs. 673-674 and 1377-1382]

- Treatment of non-traumatic hemarthrosis in Hemophilia A patient = Desmoprssin (DDAVP) or if N/A give Factor VIII 12.5 U/kg. Most patients will require **25 U/kg** every 24 hours x 2-3 days for most bleeds. If severe bleeding 50 U/kg

- Treatment of mild bleeding in von Willebrand's disease = Desmoprssin (DDAVP)

- Treatment of severe bleeding in von Willebrand's disease = Cryoprecipitate

- 1 unit of platelets = will increase platelet count 5-10,000 in adults, 20,000/20kg in PEDS
Ask for single-donor platelets when possible (contain the equivalent of 6 units of random donor platelets)

HEMATOLOGY / ONCOLOGY

- When platelet levels decrease < 20,000/uL = concerned about risk of spontaneous bleeding

- 1 unit PRBC will increase Hgb 1 gm

- PEDS FFP = 30 cc/kg

- Most common cause of acute hemolytic transfusion reactions = transfuse wrong blood due to clerical error

- Most common adverse effect from blood transfusion = febrile, non-hemolytic reaction. Reaction to anti-leukocyte and antiplatelet antibodies (prevent by using washed RBCs)

- **DIC** most helpful labs = ↑ PT, ↓ platelet count and ↓ fibrinogen

- Heparin has selective use in **DIC** when fibrin deposits and thrombosis predominate. Heparin should be considered in purpura fulminans, retained dead fetus before delivery, giant hemangioma and acute promyelocytic leukemia [10, 5th ed., pg. 1698]

- **Sickle cell patients** = functional asplenia; risk of encapsulated organisms – S. pneumo. H. influenza, Salmonella

- Major cause of **anemia worldwide** = hookworm infection (Necator americanus); infective filariform larva penetrate skin → adult worms penetrate into intestinal mucosa and feed → luminal blood loss [10, 5th ed., pg 1871]

- Most common human enzyme defect = deficiency of RBC enzyme Glucose-6-phosphate dehydrogenase (G-6-PD); affects 1/10th of world population
Oxidant stress → **hemolytic anemia,** renal failure, low platelets
Causes: infections, metabolic acidosis (DKA), exposure to oxidant drugs and ingestion of fava beans
Drugs associated = antimalarials (primaquine), nitrofurantoin, phenazopyridine (pyridium), sulfas
Self limited because only the older RBCs hemolyze [10, 5th ed., pg1676-1677]

HEMATOLOGY / ONCOLOGY

PEDIATRICS

CONGENITAL HEART DISEASE - CYANOTIC
Mnemonic - (5 Terrible T's) [ANK]
Right to left shunts causing early cyanosis

T	**T**runcus Arteriosus (single arterial trunk exits ventricular portion of heart)
T	**T**ransposition of the Great Vessels Most common cause of cyanosis or CHF within the first **3 days** of life Aorta originates from RV and pulmonary artery from the LV Systemic veins drain → RA and pulmonary vein → LV
T	**T**ricuspid Valve Atresia
T	**T**etralogy of Fallot (Most common cyanotic congenital heart disease in kids > **4y/o**)
T	**T**otal Anomalous Pulmonary Venous Return

When trying to remember the 5 T's think of counting the fingers on your hand
[ANK from Drs. Lisa McQueen, Nathan Allen]

1 finger	One arterial trunk exits the ventricle (Truncus Arteriosus)
2 fingers	Cross middle finger over index finger) = 2 vessels transposed Transposition of the Great Vessels
3 fingers	Tricuspid Valve Atresia (3 = Tri)
4 fingers	Tetralogy of Fallot (4 = Tetra)
5 fingers	Palm open all (total) fingers up = Total Anomalous Pulmonary Venous Return Also, 5 = 5 words

TETRALOGY OF FALLOT
Mnemonic - (POSH) [ANK]

P	**P**ulmonary stenosis (RV outflow tract obstruction - RVOTO)
O	**O**verriding aorta (Dextroposition of the aorta); the aortic valve is situated above the VSD with biventricular connection (connected to both the RV and LV)
S	**S**eptal defect - VSD; holosystolic murmur
H	**H**ypertrophy – RV

CXR = boot-shaped heart

PEDIATRICS

MANAGEMENT OF HYPERCYANOTIC OR TET SPELL
Mnemonic - (5P's) [ANK from Dr. Angela McCormick]

- **P**osition - Knees to chest or squatting → ↑ venous return to heart and ↑ SVR
- **P**ain control - O2 and Morphine 0.2 mg/kg SQ or IM per dose
- **P**ropranolol 0.05 to 0.1 mg/kg IV or Esmolol (with consultation)
  Relaxes infundibular muscle spasm that causes RVOTO
  or
- **P**henylephrine 10 μg/kg bolus followed by infusion 2 to 5 μg/kg/min
  → ↑ SVR → ↑ BP [12, 6th ed. pg. 761]
- **P**rostaglandins in neonates may be lifesaving = 0.1μg/kg/min [10, 5th ed. pg. 2283]

If hypercyanotic episode is not recognized and treated early it may be fatal. Other complications = seizures, cerebral thrombosis, profound lactic acidosis and cardiac dysrhythmias [12, 6th ed. pg. 761].

CONGENITAL DISORDER	CV MANIFESTATIONS
Downs	Atrial Septal Defect (ASD) SEM at left sternal border *Left to right* intra-cardiac shunt; present > 6 month = CHF
Turners	Coarctation of the aorta (↓pulse LEs, HTN UEs, SEM at cardiac base radiates → interscapula) = CHF
Rubella	Patient ductus arteriosus (PDA) *Left to right* intra-cardiac shunt; present < 6 months = CHF

CONGESTIVE HEART FAILURE
[12, 6th ed., pgs. 761-763]

Infants present with poor feeding, labored breathing and sweating,

Most common cause of CHF < **1 day** = non-cardiac (↓H/H, ↓glucose, ↓O2, ↓Ca2+, sepsis, acidosis) or premature neonate with PDA

Most common cause of cyanosis or CHF within the first **3 days of life of life** = Transposition of the Great Vessels (TGV)

Most common cause of CHF **1st week** of life in full-term newborns = Hypoplastic LV (HPLV)

Most common cause of CHF **2nd week** of life in full-term newborns = Coarctation of the aorta

VSD = CHF **4 to 12 weeks** of life, unless complicated by other cardiac disease then earlier

TREATMENT OF CHF

Lasix 1 to 2 mg/kg IV
Digoxin 0.05 mg/kg per day, in infants up to 2 y/o. Give first digitalizing dose in ED (50% daily dose), followed by one-fourth of the daily dose IV at 6 to 8 hour intervals

Cardiogenic shock → inotropic agents
Dopamine or
Dobutamine

PEDIATRICS

If Inotropic Support Fails
Combination of Nitroprusside or NTG + Dopamine

- Blue baby = Terrible T's
- Mottled or gray baby = Coarctation of the aorta or aortic stenosis
- Pink baby = VSD, PDA

COMMON CAUSES OF NEONATAL SEPSIS / MENINGITIS (< 1 MONTH)
Mnemonic: GEL [Sanford Guide 2015]

G	**G**roup B Strep. *(Strep. agalactiae)* Most common cause 49%
E	**E**. *coli* 18%
L	**L**isteria 7%

TREATMENT OF NEONATAL MENINGITIS
[Sanford Guide 2015]

Ampicillin (will cover Listeria) 50 mg/kg q 6 hrs (max dose 2 grams) +
Cefotaxime (Claforan) 200 mg/kg/day divided q 6-8 hrs (max dose 2 grams)

or Amp + Gentamycin

If CSF pleocytosis and negative gram stain consider adding Acyclovir empirically for HSV

MENINGITIS > 1 MONTH
[Sanford Guide 2015]

1) Strep. pneumoniae
2) Neisseria meningitidis
3) H. influenzae

TREATMENT OF MENINGITIS > 1 MONTH TO 50 YEARS
[Sanford Guide 2015]

- Dexamethasone + Ceftriaxone or Cefotaxime + Vancomycin
- Ceftriaxone = 100 mg/kg (2 gm IV max) q 12 hrs
- Dexamethasone = 0.15 mg/kg IV q 6 hrs x 2 to 4 days; give 15 min prior or con-comitant with first dose of antibiotic to prevent neurologic complications.
- Vancomycin = 15 mg/kg IV q 6 hours; Adults max dose of 2-3 gm/day is suggested:
- 500 to 750mg IV q 6 hours

Most common cause of aseptic meningitis in US = Enterovirus (coxsackie)

Neisseria meningitidis prophylaxis rule of 2s = 2 hours within 2 feet require prophylaxis; also if direct mucosal contact with patient's secretions (mouth-mouth, intubation or nasotracheal suctioning) [10, 5th ed., pg. 1538]

If > 50 years or alcoholism or other debilitating associated diseases or impaired cellular immunity add, Ampicillin 2gm IV q 4h to regimen above cover possible *Listeria*

PEDIATRICS

TYPICAL CSF CHARACTERISTIC OF NORMAL & INFECTED HOSTS

Case	Color	Opening Pressure	WBC	Glucose	Protein
Normal Infant	Clear	< 180 mm	< 10mm3	> 40 mg/dL	90 mg/dL
Normal child or Adult	Clear	< 180 mm	0	> 40	< 40
Bacterial Meningitis	Cloudy	> 200 mm	200-10,000 (> 80% PMN)	< 40	100 – 150
Viral Meningitis	Clear	< 180 mm	25-1,000 (< 50% PMN)	> 40	50-100
Cryptococcal Meningitis	Clear	> 200 mm	50-1,000 (< 50% PMN)	< 40	50-300

- Also, ask for Gram's Stain, Culture and Sensitivity and consider: + India Ink, cryptococcal antigen, HSV-PCR, AFB and antigen detection studies (counterimmunoelectrophoresis), latex agglutination and coagglutination are methods for detecting specific antigens

- [Clin Infect Dis. 2004;39:1267-1284] → Consider lactate; CSF lactate level > than 31.53 → suspect bacterial meningitis

- [Rev Neurol Dis. 2006;3(2):57-60] → does not recommend lactate → nonspecific

- CSF glucose is dependent on serum glucose. Rough guideline =
  CSF glucose is normally > serum glucose/2
  CSF/blood glucose ratio less or equal to 0.4 → suggests bacterial meningitis

- When peripheral cell counts are normal, CSF from traumatic LP should contain
  1 WBC per 700 RBCs [10, 5th ed., pg. 1533]

FEBRILE SEIZURES
[Dr. Collucci, 12, 6th ed, pg. 803 and PEDIATRICS Volume 121, Number 6, June 2008]

- History of febrile seizure in first degree relative = most consistently identified risk factor for febrile seizure

- 3% of all children

- 6 months to 5 years

- **Simple Febrile Seizures** = < 15 minutes, generalized tonic-clonic, no focal neuro deficits, mild post-ictal period, occurs once in 24 hours

- **Complex Febrile Seizures** = prolonged (>15 minutes), are focal, or occur > once in 24 hours

- 30% recurrence (especially if first seizure occurs in child < 1 year old)

- Risk factors for recurrence:
  - Young age of onset < 18 months old (strongest and most consistent risk factor for recurrence)
  - History of febrile seizure in first degree relative
  - Low grade temperature in ED
  - Brief duration between fever and seizure

PEDIATRICS

- Simple vs. complex is NOT predictive of the risk of recurrence

- Viral infections are common causes: influenza, adenovirus and parainfluenza. **Human herpes 6 (HHV-6)** infection is a particular risk for febrile seizures

- Risk of epilepsy = 1%, same as general population

- If onset < 12 months, multiple seizures or family hx of epilepsy – risk of epilepsy = 2.4%

- Neither a decline in IQ, academic performance or neurocognitive inattention nor behavioral abnormalities have been shown to be a consequence of recurrent simple febrile seizures

- Long-term therapy antiepileptic therapy is not recommended for simple febrile seizures

- In situations in which parental anxiety associated with febrile seizures is severe, intermittent oral diazepam at the onset of febrile illness may be effective in preventing recurrence

- Antipyretics may improve comfort of the child, they will not prevent febrile seizures

- The AAP stance on Febrile seizures from 1996
  - "Strongly recommend" lumbar puncture (LP) in patients < 12 months
  - "Consider" LP 12 to 18 months
  - LP is not routinely necessary in patients > 18 months.

STREPTOCOCCUS IDENTIFICATION

Alpha-hemolytic
- Streptococci from a GREEN zone around their colonies as a result of incomplete lysis of red blood cells (RBC's) in the agar
- *Streptococcus pneumoniae*
- *Strep viridans* (eg, *Strep mitis* and *Strep mutants*)

Beta-hemolytic
- Streptococci form a CLEAR zone around their colonies as a result of complete lysis of RBC's. Beta-hemolysis is due to the production of enzymes called hemolysins
- *Streptococcus pyogenes (Group A)*
- *Streptococcus agalactiae (Group B)*

Group (Lancefield Groups A-U)
- Streptococci are determined by antigenetic differences in C carbohydrate in the cell wall
- Group A *Streptococcus Pyogenes*
- Group B *Streptococcus agalactiae*
- Group D Enterococci (eg. *Streptococcus faecalis*)
  → causes urinary, biliary/abdominal and cardiovascular infections
- Group D Non-enterococci (eg. *Streptococcus bovis*, alpha hemolytic) → Gastric CA association
  Group D hemolytic reaction is variable. Some are beta, alpha or non-hemolytic
- Non-group = *Streptococcus pneumoniae* and *Strep viridans*
- Groups C, E, F, G, H and K-U streptococci infrequently cause human disease

PEDIATRICS

M protein
- Associated with virulence and determines the type of Group A β-hemolytic Strep. It interferes with ingestion by phagocytes

Anaerobic /microaerophilic Strep
- Peptostreptococci, variable hemolysis, cause mixed GI infections

STREPTOCOCCUS PNEUMONIAE "PNEUMOCOCCUS"
Mnemonics - (COMMONPLACES) [1]

C	**C**onjunctivitis
O	**O**titis
M	**M**edia
M	**M**eningitis
O	**O**ptochin sensitive Strep viridans eg, Strep mitis and Strep mutants – both alpha hemolytic, are NOT inhibited by **O**ptochin
N	**N**asal Sinusitis

P	**P**enicillin (Drug of choice)
L	**L**obar pneumonia
A	**A**lpha hemolytic (Non-Grouped)
C	**C**apsule
E	**E**lderly are candidates for vaccination
S	**S**putum-rusty

STREPTOCOCCUS PYOGENES
Mnemonic - (PIECES)

Group A, beta-hemolytic Strep, Gram + spherical cocci in pairs or chains

P	Pharyngitis (get circumoral pallor) Penicillin = drug of choice If PCN allergy you can use Zithromax Zithromax Dose = 12mg/kg/day x 5 days – not the traditional 10mg/kg day #1 than 5mg/kg days #2 to 5
I	Impetigo
E	Erysipelas
C	Cellulitis
E	Erythrogenic Toxin → Scarlet Fever
S	Sequelae: 1) Rheumatic Fever (most common after pharyngitis) 2) Poststreptococcal GN (most common after skin infection-impetigo)

PEDIATRICS

- Rheumatic Fever (RF) is a NON-infectious autoimmune disease typically occurring 2-4 weeks after Group A Strep pharyngitis. The immunologic reaction results from cross-reactions between streptococcal antigens and antigens of joint and heart tissue

- RF is prevented if strep infection treated within 8 days after onset

- Early antibiotic treatment = earlier resolution of symptoms & shortens course of illness by 1 day [10, 5th ed., pg. 969-972]

- Not common < 2 y/o

- Occurs winter, spring

- Responsible for < 15% of pharyngitis in patients > 15 years old [10, 5th ed., pg. 969-972]

- Rapid Strep sensitivity = 60 – 95% [10, 5th ed., pg. 969-972]

- Acute glomerulonephritis (AGN) = HTN, edema face/ankles and "smoky" urine. May be prevented if strep treated early, however cannot be prevented with PCN after onset of symptoms

- Reinfection with strep rarely leads to recurrence of AGN

- "Doughnut lesions" = erythematous papules with a pale center located on both the soft and hard palates, pathognomonic for Group A Strep pharyngitis

5 MAJOR MODIFIED JONES CRITERIA FOR RHEUMATIC FEVER
Mnemonic - (EM Physicians Can Snuggle Continuously) [7, with modifications]

EM	**E**rythema **M**arginatum (pink rings on the trunk and inner surfaces of the arms and legs)
Physicians	**P**olyarthritis (symmetric, migratory-see mnemonic on migratory arthritis)
CAn	**CA**rditis (steroids may be helpful)
SNuggle	**S**ubcutaneous **N**odules
COntinuously	**Ch****O**rea (Sydenham's chorea = "Saint Vitus' Dance")

Clinical diagnosis of Rheumatic Fever is made if 2 Major or 1 Major and 2 Minor criteria are present in patient with preceding Strep infection evidenced by:

1) ↑ ASO titer
2) Positive throat culture
3) Recent Scarlet Fever

Minor Criteria
- Fever
- Arthralgia
- ↑ PR interval
- ↑ ESR or CRP
- Previous rheumatic fever

PEDIATRICS

CAUSES OF MIGRATORY ARTHRITIS
Mnemonic - (RF HSP LSD)

A tough one to remember, Try "if you had rheumatic fever, or HSP, you would want LSD."

RF	**R**heumatic **F**ever

H	**H**SP
S	**S**epsis (Strep, Staph, GC, Meningococcal)
P	**P**ulmonary Infection (Mycoplasma, Histo, Coccidia)

L	**L**yme disease
S	**S**BEndocarditis
D	**D**rugs (Ceclor)

SINUSES PRESENT AT BIRTH
Mnemonic - (ME)

M	**M**axillary
E	**E**thmoid

Fyi: adult patients →
Leaning forward exacerbates maxillary sinusitis
Supine position exacerbates ehtmoid sinusitis.
CT scan is the gold standard for diagnosis of sinusitis.

CONGENITAL TOXOPLASMOSIS (TOXOPLASMA GONDII)
Mnemonic - (THC3) [1]

T	**T**oxoplasmosis
H	**H**ydrocephalus
C	**C**horioretinitis **C**erebral Calcification **C**at feces and raw meat sources

Note: The acquired form in adults may present as a mononucleosis-like syndrome.

DIAGNOSTIC CRITERIA FOR KAWASAKI SYNDROME
- Fever for at least 5 days duration
- Illness that is not explained by other known disease process

PEDIATRICS

AND 4 of the following 5 For Diagnosing Kawasaki Syndrome
1. Bilateral conjunctivitis
2. Changes of lips and oral mucosa
   (strawberry tongue, red/fissured lips, oropharyngeal edema)
3. Changes of the extremities
   (erythema of the palms and soles, edema of the hands and feet, periungual desquamation)
4. Polymorphous rash
5. Cervical lymphadenopathy

Complication = coronary artery aneurysms; high suspicion tx =Gamma-globulin, high dose ASA

REYE SYNDROME
- ↓Glucose
- ↑LFT's, except normal BILI, ↑ ammonia
- Mental Status changes with ↑ICP and seizures

DIPHTHERIA
Corynebacterium diphtheriae, club-shaped G+ aerobic rod, no capsule
Spread person-person nasopharyngeal secretions (place patient in respiratory droplet isolation)
Gray-green pseudomembrane, do not remove → ↑bleeding; "bull neck" appearance
Exotoxin ¬→ disrupts protein synthesis → leads to cardiac and CNS disease
Myocarditis → cardiomyopathy, CHF, Dysrhythmias
CNS & PNS myelin sheath deterioration → peripheral neuropathy, muscle weakness (palate 1st)
Diagnosis: by culture of pharynx and nose (alert lab as special media is needed)

Diphtheria Treatment
Active immunization +
Antitoxin (dose depends on site of infection and duration of symptoms) +
Erythromycin or Penicillin x 7-14 days [Sanford, 2015]

CHILD ABUSE
1. Neglect   52%
2. Physical   24%
3. Sexual   12%
4. Emotional 6%

- Greatest risk factor = lower socioeconomic status
- Head trauma is the most common cause of mortality
- Shaken Baby / Shaken-Impact Syndrome = SDH, retinal hemorrhages, rib fractures, metaphyseal fractures of long bones

PEDIATRICS

HENOCH-SCHÖNLEIN PURPURA PEARLS

Mnemonic - (PANDAS) [ANK, mnemonic provided by Dr. Collucci] [12, 6th ed, pg. 886]

P	**P**urpura, palpable – blanching, buttocks and lower legs
A	**A**bdominal pain 85%, nausea & vomiting (may have → melena/hematochezia; 8% massive GI bleed or intussusception)
N	**N**ephritis 25 – 50%
D	**D**iarrhea
A	**A**rthritis, polymigratory 60 – 80%
S	**S**crotal edema (2 – 35%) can mimic torsion

- IgA dominant immune complexes → systemic vasculitis - mainly arterioles and capillaries
- 75% cases 2 to 4 years
- Typically follows URI, in spring
- Intussusception = ileal-ileal secondary to small bowel vasculitis
- Corticosteroids → improvement in GI, Renal and CNS complications

MECKEL'S DIVERTICULUM – RULE OF 2'S

- TWO % of population is affected (of those patients 96% never have problems)
- TWO years of age = most will have presented
- TWO feet from terminal ileum = diverticulum location
- TWO x more common in male > female

- Most common cause of massive rectal bleeding = Meckel's (painless, brick or bright red bleeding)
- Most common cause of minor rectal bleeding (peds) = Anal fissures (painful)
- Child may be well appearing, or may have bilious vomiting & abdominal distention with rectal bleeding
- The bleeding occurs from ulcers in the diverticulum secondary to ectopic gastric mucosa → which is what takes up the Technetium (85% sensitivity and 95% specificity)
- May cause intussusception
- Definitive treatment = surgery

RECTAL PROLAPSE (PROCIDENTIA)

Disease of the extremes of age; PEDS think Cystic Fibrosis; boys > girls;
Elderly = history of excessive staining; women > men

GASTROSCHISIS AND OMPHALOCELE
[10, 5th ed, pgs. 103]

- Gastroschisis = defect in abdominal wall → evisceration of abdominal structures without a sac
- Omphalocele = defect in umbilical ring → intestines protrude in a sac
- Gastroschisis 2x more common than Omphalocele

ED Management = NGT and place plastic covering to prevent heat and water loss.

PEDIATRICS

ANAL PRURITUS

- Most common cause of anal pruritus in kids = Pinworms (Enterobius vermicularis) [12, pg. 487]
- Most prevalent parasite in the US = Enterobius vermicularis (pinworm); 20 to 30% kids infected [10, 5th, pg. 1318]
- Most common cause of perianal cellulitis in young children = *group A Streptococcus*

HIRSCHSPRUNG'S DISEASE
[10, 5th ed, pgs. 2305-2306]

- Most common cause of obstruction in newborn (functional obstruction)
- 75% recto-sigmoid involvement
- Diagnosis made in nursery as there is no passage of meconium for 24-48 hours
- Bilious vomiting, tarry diarrhea, distended abdomen, poor feeding
- 5x more common in boys
- Associated with Down Syndrome
- Diagnosis
    - Suspected on barium enema - normal segment of colon with proximal dilation
    - Confirmed by biopsy – congenital absence of parasympathetic ganglion cells
- Complication: Toxic Megacolon = progressive enlargement of proximal segment
- Complication: Enterocolitis (abdominal distension, fever, ↑WBC and bloody stool)
- Treatment is surgical; NGT, rectal tube

INTUSSUSCEPTION

- Most common cause of obstruction in children < 3 y/o; male > female
- Siblings of affected patients have a relative risk 20x > general population
- Most common site ileo-cecum – 80%
- Causes: Meckel's Diverticulum, inverted appendix, tumors, polyps, duplication, lymphoma, HSP (if HSP site = ileal-ileal)
- 50% preceded by viral illness
- Colicky abdominal pain (acute pain, patient draws up legs)
- **Bilious** vomiting
- Check for "currant jelly" stools (late finding) only present in about 20% of cases
- Dance's Sign = RUQ sausage shaped mass (intussusception) & empty space in RLQ (cecum → RUQ)
- Diagnosis & Treatment = Air contrast or Barium enema; notify Peds surgery before enema (Air Contrast Enema advantages: better control of colonic pressure used for the reduction, safer, faster, less expensive and more effective than barium enema
- Complications = 5 to 10% re-intussusception within 48 hours; severe sepsis or septic shock; bowel perforation

MALROTATION

- Acute bilious emesis, poor feeding, irritability and bloody stool in neonate
- Volvulus is a complication of malrotation
- Duodenal obstruction without volvulus
- Diagnosis: Upper GI with small bowel follow-through ("cork-screw sign")
- Surgery is the only definitive treatment

PEDIATRICS

HYPERTROPHIC PYLORIC STENOSIS

- 95% of cases between 3-12 weeks; rare before 1 week and after 3 months
- Male > female
- More frequently in first born males
- Familial incidence in 50%
- NON-bilious projectile vomiting just after feeding
- ↓Cl-, ↓K+; hypo-chloremic, hypo-kalemic metabolic alkalosis
- "Olive" in RUQ /pyloric tumor is pathognomonic – 70 to 80% of patients
- Diagnosis: Ultrasound is the gold standard; if US N/A then UGI
- Treatment: fluid, correct electrolytes and surgery

PEDIATRIC PEARLS

- Most common cause of croup = Parainfluenza virus; usually < 3 y/o

- "Previous" most common cause of epiglottitis = H. influenzae now Strep Pneumo & GABHS; 3 to 6 y/o

- Most common cause of bronchiolitis = RSV

- Most common serious bacterial illness (SBI), causes 5% of fever without a source = UTI

- Most common cause of infantile diarrhea = Rotavirus (ROTA = Right Out The Anus) [17]

- Neonatal conjunctivitis (ophthalmia neonatorum): First month of life
  Day 1-2 = chemical, Day 3-5 Gonorrhea, Day 6 to first month of life = Chlamydia trachomatis

- Most common viral cause of conjunctivitis = Adenovirus

- Most common viral cause of otitis media = RSV

- Most common bacterial cause of otitis media = Strep pneumoniae; other two = Haemophilus and Moraxella. Treat all with High Dose Amoxicillin 90 mg/kg/day div q 12 or q 8 hours

- OM best confirmed = decreased mobility of TM and loss of normal landmarks

- Most common complication of otitis media = hearing (conductive) loss [10, 6th ed., pg. 930]
  Another common complication of otitis media = perforation, which usually heals within 7 days

- Most common intracranial complication of otitis media = meningitis

- Most common bacterial cause of pharyngitis = Strep pyogenes (Group A, beta hemolytic strep)

- Monospot is positive in 30% of children 0 to 20 months with mononucleosis; Monospot may be negative the first week of illness [10, 5th ed., pg 971]

- Most common cause of dysuria in school age girls = non-specific vulvovaginitis

- Most common manifestation of GC infection in children = vaginitis. Obtain specimen from vaginal introitus [10, 5th ed., pg. 1395]

PEDIATRICS

- Roseola infantum (exanthem subitum) → human herpes 6-infection → febrile 3-5 days →↓ fever →↑ macular rash. Most common exanthem in children < 2 year of age

- Rubella (German measles) → maculopapular rash spreads in a centrifugal pattern (head to feet); lymphadenopathy typical → postauricular and occipital; "blueberry muffin" skin rash
  Rubella and EBV are the most common viral cause of arthritis

- Most common cause of cataracts in newborn = Rubella (German measles)

- Rubeola (Measles) → maculopapular rash spreads in a centrifugal pattern (head to feet); 3-4 day prodrome = C x 3 = Cough, Coryza, Conjunctivitis and Koplik's spots (white lesions on buccal mucosa = pathognomonic)

- Erythema Infectiosum → fifth disease → parovirus B 19 → "slapped cheek" disease

- Most common cause of acute ataxia is = post-infection (especially varicella); r/o tox, cerebellar tumors

- Mumps = 5-15 y/o; infective 3 days before → 7 days after salivary gland swelling; 70-80% bilateral; Spread = respiratory droplets.
  Complications: meningitis. encephalitis, orchitis 15-25% postpubertal men (sterility uncommon; 70% unilateral), uncommon = GBS, transverse myelitis, oophoritis, mastitis, myocarditis, pancreatitis

- Most common growth plate fracture = Salter-Harris type II (physis and metaphysis) 75% [12, pg. 674]

- Most common Salter-Harris fracture likely to result in bone-growth arrest = type V [27, Vol. 14, No. 9, pg. 111]

- Most common cause of hip pain in children < 10 y/o = transient tenosynovitis; ↑ in boys; follows viral illness

- Most common elbow fracture in childhood = supracondylar fracture of the distal humeral metaphysis – the distal fragment is most commonly displaced posteriorly
  Complication = Volkman's ischemic contracture (compartment syndrome) [12, pg. 677, 1230]

- Legg-Calve-Perthes disease = avascular necrosis of pediatric femoral head; age 2-10y/o; 5x ↑ in boys; b/l 15%

- Hip pain → knee in adolescent obese patient → SCFE (slipped capital femoral epiphysis); 2.5x ↑ in boys; ORIF

- Nursemaid's elbow common 1 to 5 years old – stretch annular ligament and subluxation of radial head; Reduce: downward pressure on child's radial head by the docs thumb, supinate and flex elbow

- Greenstick fracture = incomplete, angulated fractures of long bones; may need to complete fracture to achieve anatomic reduction [10, 5th ed., pg. 169]

- Most common fractured bone in children = clavicle

- Fractures suspicious of abuse: spiral fractures of long bones, metaphyseal chip fractures

- Maximum dose of lidocaine of infiltration without epi = 5 mg/kg and with epi = 7 mg/kg

PEDIATRICS

- Codeine dose = 0.5 to 1.0 mg/kg (Acetaminophen with codeine = 120 mg/12mg per 5cc)

- Most common Chronic Disease of childhood = Asthma [10, 5th ed., pg. 939]

- Most common cause of home accidental death < 6 y/o = FB aspiration (usually in right mainstem bronchus)

- Age Group most likely to aspirate a FB = 1 to 3 years with peak incidence at age 2 [10, 5th ed., pgs. 757-758]

- Most common foods aspirated = nuts [10, 5th ed., pgs. 757-758]

- Common cause of pediatric fatal aspiration = food (especially hot dogs) and toy balloons

- Most common Esophageal impaction → kids = coins; adults = food (meat/bones)

- If button battery has passed esophagus and patient is asymptomatic → no retrieval. If cell has not passed pylorus in 48 hours = endoscopic retrieval

- Most common cause of metabolic acidosis in kids = prolonged diarrhea

- Congenital adrenal hyperplasia = adrenal insufficiency → ↓Aldosterone →↓Na+ more common > K+ and ↓ cortisol; Refractory ↓BP: Girls = ambiguous genitalia (look both male and female) → treatment = Hydrocortisone 2 mg/kg IV/IO bolus (max 100mg) [PALS provider manual]

- Most common cause of death 1 month to 1 year = SIDS (RARE before 1 month, risk factors for SIDS = winter, male > female if infectious etiology otherwise, male=female, if mom < 20 y/o, smokes, drugs, no pre-natal care and lower socioeconomic group) [12, pg. 599]

- Pott's puffy tumor = complications of frontal sinusitis → subperiosteal abscess with soft tissue swelling causes pitting edema over the frontal bone; Although it can affect all ages, it is mostly found among teenagers and adolescents

- Most common pediatric dysrhythmia = PSVT heart rate usually > 220, in adults less
Treatment = Adenosine (Adenocard) 0.1 mg/kg (max 6mg) → 0.2 mg/kg (max 12 mg), may repeat x 1
SVT Cardioversion = 0.5 to 1.0 J/kg → 2 J/kg [2010 AHA Guidelines for Cardiopulmonary Resuscitation and Em Cardiovascular Care, S712]

- Verapamil is contraindicated in infants

- Most common cause of cardiac arrest = respiratory arrest

- Most common rhythm disturbance in pediatric arrest = bradycardia

- Bradycardia is most commonly an indicator of hypoxemia in the newborn

- First line drug for bradycardic arrest = Epinephrine 0.01 mg/kg of 1:100,000 IV/IO; ETT = 0.1 mg/kg 1:1,000
Repeat Epi every 3 to 5 minutes
If ↑ vagal tone or 10 AV Block → Atropine 0.02 mg/kg (minimum dose 0.1mg; max total dose for child = 1 mg) → consider pacing [PALS 2006 provider manual, pg 123]

PEDIATRICS

- Vfib/ Pulseless VT = 2 J/kg Biphasic → CPR 5 cycles → 4 J/kg biphasic [2010 American Heart Association Guidelines for Cardiopulmonary Resuscitation and Emergency Cardiovascular Care; S708]
  Epi 0.01 mg/kg; Amio 5mg/kg

- Most common category of shock in the pediatric population = hypovolemic; fluid = 20cc/kg 0.9 NS or LR

- Peds Post-arrest - the drug of choice in treating hypotension → epinephrine infusion [10, 5th ed, pg. 98]

- ETT size in PEDS can be estimated = 16 + (age in years) / 4 or age/4 + 4
  ETT x 2 = NG/Foley catheter size
  ETT x 3 = Depth of ETT insertion
  ETT x 4 = Chest Tube size

- < 2y/o ETT size = size of patients small finger (fifth digit!)

- Position of the ET tube at the lips (in cm's) should = 3 x size of ETT

- Uncuffed ETT in children under 6-8 years of age

- Pretreatment with atropine in RSI for children < 10 years old has fallen out of favor (lack of evidence to support)

- Treatment of hypoglycemia
  Neonate = D10 5 to 10 mL/kg
  Child = D25 2 to 4 mL/kg

- Upper limit of SBP = (Age x 2) + 80

- Weight in kilograms = (Age x 2) + 8

  Newborn = 3 kg
  1 y/o = 10 kg
  5 y/o = 20 kg
  10 y/o = 30 kg

APGAR SCORE
1 and 5 Minutes after birth. Score 0-2

		0	1	2
A	**A**ppearance (color)	blue /pale	acrocyanosis	pink
P	**P**ulse	Absent	<100 beats	>100
G	**G**rimace (reflex)	no response	grimace	cough/active withdrawal
A	**A**ctivity (muscle tone)	limp	some flexion	active motion
R	**R**espirations	absent	weak cry/hypovent.	good, crying

Want score at least 7

PEDIATRICS

SURGERY / GI / TRAUMA

SMALL BOWEL OBSTRUCTION - HISTORY
Mnemonic - (VODKA) [6]

V	Vomiting
O	*Obstipation - Old Scar
D	Distension
K	(K) Crampy
A	Abdominal Pain

*Obstipation = 1) the act or condition of obstructing 2) extreme constipation due to obstruction

CAUSES OF SMALL BOWEL OBSTRUCTION
Mnemonic - (BEN VIP) [3, with modification]

B	**B**ands (adhesions) = most common cause of SBO; from previous Sx or Crohn's
E	**E**xternal Hernias = 2nd most common cause of SBO
N	**N**eoplasm (note: neoplasm is the most common cause of LBO)

V	**V**olvulus
I	**I**ntussusception
P	**P**ackets (swallowed FBs, bezoars, etc)

SBO PEARLS
[10, 5th ed. pg. 1283, 1284 & 1287]

SBO → ↑ intraluminal pressure → capillary and lymphatic obstruction → ↑ bowel wall edema
Aggressive treatment has ↓ Mortality Rate from 60% in 1900 to less than 5% today
35 to 50% of patients with complete SBO → resolution without surgical intervention

CAUSES OF LARGE BOWEL OBSTRUCTION
Mnemonic - (CAtch VD)

CAtch	**CA**ncer = most common cause (> 65%) Most common cancer = adenocarcinoma
V	**V**olvulus = 3rd most common cause (5%) Most common site = sigmoid (70%) and cecum (30%)
D	**D**iverticulitis = 2nd most common cause (20%), secondary to scarring

SURGERY / GI / TRAUMA

- Other causes of LBO:
  Inflammatory disorders, benign tumors, foreign bodies, radiation and fecal impaction
- Ogilvie syndrome = enormous dilation of the RIGHT colon without mechanical obstruction (pseudo obstruction)
- History of LBO = diffuse colicky pain, obstipation...vomiting LATE or ABSENT; distension most common and prominent physical finding

CECAL VOLVULUS
[10, 5th ed., pg. 1335]

- Occurs in all ages, but most common in 25 to 35 years of age
- Not associated with chronic constipation (unlike Sigmoid)
- Onset of pain = acute; in sigmoid volvulus = more gradual
- Treatment = surgery, non-op decompression often unsuccessful

Treatment of non-strangulated Sigmoid Volvulus = decompression and detorsion using a rectal tube via the sigmoidoscope = 85-95% success [10, 5th ed., pg. 1334]

CAUSES OF ILEUS
Mnemonic - (Nurses (RN) and Physician Assistants (PA) FiX MI Burn GAP)

R	**R**etroperitoneal hematoma
N	**N**ephrolithiasis

P	**P**yelonephritis or Pneumonia
A	**A**bdominal Surgery

FiX	**F**x (fractures – lumbar, rib)
MI	**M**yocardial Infarction
Burn	**Burn**s, especially if > 20%

G	**G**allstone ileus or Gastroenteritis
A	**A**ppendicitis
P	**P**ancreatitis

ILEUS	SBO
Minimal abdominal pain	Crampy abdominal pain
↓ or absent bowel sounds	↑ or normal bowel sounds
Gas in SI & colon on x-ray	Gas in SI only on x-ray
Nausea & vomiting	Nausea & vomiting
Obstipation & failure to pass flatus	Obstipation & failure to pass flatus
Abdominal distention	Abdominal distention

SURGERY / GI / TRAUMA

APPENDICITIS
Mnemonic - (PAVEL'S P'S) [1, with modifications]

P	**P**ain-periumbilical → RLQ
A	**A**norexia
V	**V**omiting
E	**E**levated temperature, mild/normal. If elevated, think perforation
L	**L**eukocytosis – may be normal
S	**S**igns [10, 5th ed, pg 1294] Rovsing's = palpate LLQ → tender RLQ Obturator = patient supine, flex and internally rotate right hip → tender RLQ Iliopsoas = patient asked to extend right hip → tender RLQ

P	1) Perforation (high rate in kids, elderly and pregnancy) 2) Phlegmon (suppurative inflammation of SQ connective tissue) 3) Periappendiceal abscess 4) Peritonitis

Pathophysiology = lumen obstruction (fecaliths –most common, lymphoid hyperplasia, dietary matter, worms, tumors, granulomatous disease, adhesions) → increased edema as mucus continues to be secreted → edema and vascular compromise → followed by bacterial invasion

Most common cause of surgical abdomen in children = appendicitis
> 66% cases are in patients < 30y/o (peak incidence = 11 to 20 y/o)

Condition most commonly mistaken for acute appendicitis in children = mesenteric adenitis and gastroenteritis

Most common surgical emergency in pregnancy = appendicitis
Higher perforation rate in pregnancy [10, 5th ed, pgs 2425-2427]
Fetal loss 20% with perforated appy
Pain may be in RUQ [10, 5th ed, pgs 2425-2427]
Most patients with appendicitis will have pain in RLQ, even in 3rd Trimester [Am J of Obst and Gyne 2000]

Threshold for human teratogenesis = 10 rad; fetus most vulnerable 8 to 15 weeks gestation
CT abdomen = 3.5 rad [12, 6th ed., pg. 675]

1 rad (Radiation Absorbed Dose) = 10 mgray (Gy); Gray = the SI unit
Gray (Gy) = amount of radiation required to deposit 1 joule of energy in 1 kg of any kind of matter

1 rad = 0.01 sievert (Sv) = 10 mSv

Sievert (Sv) = a unit used to derive a quantity called dose equivalent. This relates the absorbed dose in human tissue to the effective biological damage of the radiation. Not all radiation has the same biological effect, even for the same amount of absorbed dose. [www.bt.cdc.gov/radiation/glossary.asp]
1 sievert (Sv) = 100 rem (Roentgen Equivalent in Man)

SURGERY / GI / TRAUMA

CAUSES OF PANCREATITIS
Mnemonic - (ABCDEF SHIP LIST) [1, with modifications]

A	Alcohol Alcohol and biliary make up 85% of cases
B	Biliary (Gall Stones)
C	Cancer – pancreatic
D	Drugs (see below)
E	ERCP (3% post procedure)
F	Familial

S	Scorpion stings
H	↑Hyper-calcemia (rare)
I	Idiopathic = 3rd most common cause (thought to be a form of microlithiasis)
P	Posterior duodenal ulcer erosion
P	Pancreas divisum

L	Lipids ↑TG > 1,000 mg/dL
I	Infection (mumps, coxsackie, HBV, EBV, influenza, legionella, mycoplasma, West Nile Virus)
S	Surgery – post op
T	Trauma

DRUGS THAT CAUSE PANCREATITIS
Mnemonic - (PAST DATE) 2

P	Pentamidine	Propofol
A	Azathioprine (Imuran) common cause	Amiodarone
S	Sulfa	Steroids
T	Thiazides	Tylenol [12, pg. 788]

D	Depakote (valproic acid)	Diphenoxylate (lomotil)
A	ASA and NSAIDs	Amlodipine
T	Tetracyclines	Tamoxifen
E	Ethacrynic acid	Estrogen

HIV medications → all 4 are nucleoside reverse transcriptase inhibitors (NRTIs), inhibit viral reverse transcriptase, thus preventing / interfering with the production of a DNA copy of viral RNA

Didanosine (ddI) (also 15% neuropathy)
Zalcitabine (ddC) (also 30% neuropathy)
Stavudine (d4T) (also 30% neuropathy)
Lamivudine (Epivir)

SURGERY / GI / TRAUMA

PANCREATITITIS - RANSON'S CRITERIA
Mnemonic - (**A**ll **W**ild **G**irls **L**ike **S**occer) [ANK]

On admission

All	Age > 55
Wild	WBC > 16,000
Girls	Glucose > 200
Like	LDH > 350
Soccer	SGOT (AST) > 250

Within 48 hours
- Hct ↓ > 10%
- BUN ↑ > 5
- Ca2+ < 8
- PO2 < 60
- Base deficit > 4
- Fluid sequestration > 6l

"Although they serve as a reminder of features that portend a worse prognosis, the Ranson Criteria have poor predictive value in the acute setting that does not improve on clinical judgment" [12, 6th ed. pg. 575]

- Pathophysiology: Inappropriate activation of trypsin → pancreatic autodigestion and ↑local inflammatory mediators are released → which cause distant extra-pancreatic dysfunction
- Amylase rises within 6 to 24 hours however returns to normal in 3 to 7 days
- Lipase rises within 4 to 8 hours and stays elevated for 7 to 14 days
- ↑lipase > 3x upper limit of normal with history c/w pancreatitis to make the diagnosis
- The absolute level of serum amylase and lipase do not correlate with for disease severity and have no prognostic value [12, 6th ed. pg. 575]
- Lipase and amylase both exist in other tissues
- Experts recommend lipase over amylase when seeking diagnosis of pancreatitis [10, 5th ed. pg. 1276] and [12, 6th ed. pg. 575]
- Overall mortality rate (MR) = 5%
- Mortality rate of severe pancreatitis (see complications below) = 14 - 25% [Lancet. Jan 12 2008;371(9607):143-52; World J Gastroenterol. Mar 28 2009;15(12):1427-30]

Complications of acute pancreatitis: pleural effusions (left), ARDS, myocardial depression, DIC, renal failure, shock, phlegmons, abscesses, pseudocysts and necrosis

Signs of Hemorrhagic Pancreatitis [12, pg. 513] Not common (3%), but if present MR = 37%

Cullen sign = periumbilical ecchymosis
Turner sign = flank ecchymosis

Treatment: aggressive fluid management, analgesia, oxygen administration and early nutrition

SURGERY / GI / TRAUMA

GALLSTONES
Mnemonic - (4 F's) [ANK]

F	**F**emale
F	**F**at
F	> **F**orty
F	**F**ertile

- Gall stones < 20% picked up on KUB, cholesterol GS = 80% and these are radiolucent
- Kidney stones > 90% picked up on KUB

Charcot's Triad: present in 50 to 70% patients with cholangitis
1) Fever
2) RUQ pain
3) Jaundice

Reynolds's Pentad: Charcot's triad plus
4) Mental status changes
5) Hypotension

- Most common organisms *E. coli* (27%), Klebsiella species (16%), Enterococcus species (15%), *Streptococcus* species (8%), Enterobacter species (7%) and P. aeruginosa (7%) [http://emedicine.medscape.com/article/774245-overview]
- Murphy's Sign = inspiratory arrest, due to pain, with palpation of RUQ / over GB
- Tender RUQ → radiates to the right scapula

CHOLECYSTITIS
Labs = leukocytosis with or without a left shift, ↑ or normal aminotransferases; bilirubin usually within normal limits [10, 6th ed., pgs. 1491-1492]

- 5 to 10% acalculous
- Most common surgical emergency in elderly with abd. pain = Acute Cholecystitis [12 6th ed., pg. 503]
- Bilirubin will be elevated in cholangitis and in general have a higher fever and appear more ill then patient with cholecystitis
- Ultrasound is the procedure of choice for investigating the gallbladder [10, 5th ed., pg. 1266]
- Most sensitive and specific test for cholecystitis = nuclear scintigraphy (HIDA scan) [10, 5th ed., pg. 1267]

Cholelithiasis = gallstones in the gallbladder
Biliary colic = pain caused by a stone temporarily obstructing the cystic duct
Cholecystitis = inflammation of the gallbladder from obstruction of the cystic duct
Choledocholithiasis = stone in the common bile duct – often significant ↑Alk phos and transaminases
Cholangitis = bacterial infection superimposed on an obstruction of the biliary tree gallstone, neoplasm or stricture

PORCELAIN GALLBLADDER
- Uncommon manifestation of chronic cholecystitis, characterized by intramural calcification of the gallbladder wall
- The diagnosis is suggested by an abdominal radiograph revealing an incidental calcified gallbladder.
- Patients with a porcelain gallbladder are often asymptomatic, but are at increased risk for the development of gallbladder carcinoma (poor prognosis)

SURGERY / GI / TRAUMA

HEPATITIS PEARLS
[10, 6th edn, pgs. 1402-3 and 5th ed. pg. 1251] and [12, 6th edn, pgs. 568, 990]

Incubation period
HAV is 15-45 days
HBV is 60-90 days
HCV is 30-90 days

- ETOH Hepatitis → AST (GOT) > 2 x ALT (GPT); ratio AST/ALT > 2
- Viral Hepatitis → ALT generally >>AST; both 10 to 100x normal
- Scleral icterus when serum BILI > 2.5 mg/dl
- ↑ PT = not good, clue to complicated course; PT reflects hepatic synthetic function

HAV = RNA virus; spread = fecal → oral route
Fecal excretion of HAV usually occurs prior to symptoms of acute HAV infection
IgM antibody to HAV → indicates acute infection; no chronic carrier state

HBV = DNA virus, transmitted hematogenously and sexually
HB cAb-IgM = antibody to core antigen (cAg) → indicates acute infection
HB cAb-IgG = antibody to cAg → prior infection
HB sAb = antibody to surface antigen (sAg) → indicates acute or prior infection or immunization
HB eAb = antibody to eAg → resolving infection and decreased infectivity

HB sAg surface antigen indicates acute or chronic infection; measurable before clinical illness
HB eAg antigen associated with active acute or chronic infection, and indicates high infectivity
HB eAg comes from the core gene and is a marker of active viral replication

HCV = RNA virus
Anti-HCV = antibody that defines infection with HCV, acute or past
Risk of HCV = 0.03% per unit of blood transfused

50% HCV develop chronic hepatitis, 20% of this group develop cirrhosis within 10 years
10% of adults and 90% of neonates infected with HBV develop chronic hepatitis
In addition to chronic hepatitis and cirrhosis, HBV associated with hepatocellular CA

HEPATITIS PROPHYLAXIS

Hepatitis A = Immune Serum Globulin (ISG) = 0.02 ml/kg IM within 14 days of exposure Vaccine is available [10, 5th ed., pg. 1255]

Hepatitis B if previously unvaccinated = HBIG 0.06 ml/kg IM, simultaneously with HBV vaccine series with first shot in the deltoid [10, 5th ed. pg. 1256]

SURGERY / GI / TRAUMA

HEPATOTOXINS
→ **hepatocellular necrosis** [12, 6th edn, pgs. 568]

Acetominophen	NSAIDS
Amphotericin	Ketoconazole [10, 5th ed. pg. 1261]
INH	Amiodarone [10, 5th ed. pg. 1261]
Phenytoin	Valproic acid
Iron	Halothane
Cocaine	Ecstasy (MDMA)
Mushrooms	Carbon tetrachloride
White Phosphorus	Inorganic arsenicals, thallium & borates

Reye Syndrome: resembles fulminant liver failure, but microvesicular fatty infiltration occurs without hepatocellular necrosis

Cholestatic picture = chlorpromazine, haldol, anabolic or oral contraceptive steroids and erythromycin estolate [10, 5th ed. pg. 1261]

LIVER ABSCESS

Usually polymicrobial, *Escherichia coli* and *Klebsiella pneumonia* = 2 most frequently isolated pathogens; other pathogens = anaerobes, *Bacteroides, Fusobacterium*, microaerophilic/anaerobic Strep (*Peptostreptococcus*); aerobic Strep = *Strep. faecalis*, and *Pseudomonas*
A colonic source usually initial source of infection (diverticulits > biliary disease > appendicitis)
If hematagenous spread = *Staphylococcus aureus*
Patients with Crohn's disease = *Staphylococcus milleri*
Most common test question = *Entamoeba histolytica* → causes 10% of liver abscess cases

CHILD-PUGH CLASSIFICATION OF SEVERITY OF LIVER DISEASE
Mnemonic - (I BEAN) [?!]

	Criteria	1 Point	2 Points	3 Points
I	INR	<1.7	1.7-2.3	>2.3

	Criteria	1 Point	2 Points	3 Points
B	Bilirubin (g/dl)	< 2.0	2.0 – 3.0	> 3.0
E	Encephalopathy	None	Grade 1-2	Grade 3-4
A	Ascites	None	Slight	Moderate
N	Nutrition = Albumin, (g/dl)	> 3.5	2.8 - 3.5	< 2.8

Score of
5-6 = grade A (well-compensated disease)
7-9 = grade B (significant functional compromise)
10-15 = grade C (decompensated disease)

These grades correlate with one- and two-year patient survival:
Grade A - 100 and 85%

SURGERY / GI / TRAUMA

Grade B - 80 and 60%
Grade C - 45 and 35%

CAUSES OF FECAL LEUKOCYTES
Mnemonic - (I Can SEE leukocytes In Your Shample) [18, with modifications]

I	**I**schemic colitis

Can	**Cam**pylobacter (Guillain Barre Syndrome; poultry, eggs)

S	**S**almonella (rose spots, fever with relative bradycardia; sickle cell or asplenic patients; pet turtle or inguana; after eating poultry, eggs)
E	**E**. *coli* O157:H7 (EHEC = enterohemorrhagic E.coli) HUS → microthrombi → RF
E	**E**. *coli* (EIEC = enteroinvasive E. coli)

LEUKOCYTES

In	**IN**flammatory bowel disease (IBD)
Your	**YER**sinia enterocolitica (mimics appendicitis; erythema nodosum)
Shample	**SH**igella (seizures, Reiter's syndrome and HUS; mucoid bloody diarrhea + ↑ Fever) Reiter syndrome = can't see, can't pee, can't climb a tree = urethritis, iritis, arthritis

- Organisms that produce variable findings on microscopic stool examination, depending on the invasive properties of the strain and the degree of colonic involvement

  - *Clostridium difficile*
  - *Aeromonas*
  - *Vibrio parahaemolyticus* (most common cause of gastroenteritis in Japan)

- Viruses cause the majority of infectious diarrheas followed by bacteria [12, 6th ed., 556]

- Suggests treating diarrheal illnesses regardless of whether or not diarrhea is invasive, therefore "ascertaining the presence or absence of fecal leukocytes is superfluous" [12, 6th ed., pg. 555]

NON-VIRAL DIARRHEA PEARLS

- Most common cause of acute food poisoning in US = *Clostridium perfinges* [10, 5th ed, pgs. 1308]

- From CDC → *Campylobacter* is one of the most common causes of diarrheal illness in the US

- *Clostridium perfringens* = Patients ingest heat-resistant spores which produce an enterotoxin in the GI tract (non-invasive/non-bloody diarrhea; poultry, meat, eggs)

SURGERY / GI / TRAUMA

DIARRHEA PEARLS

- Most common cause of travelers diarrhea = *Enterotoxigenic E. coli* (ETEC) produces a toxin that acts on the intestinal lining (non-invasive / non-bloody diarrhea)

- *Enteropathogenic E. coli* (EPEC) attach to the intestinal mucosa, causing diarrhea in children and adults → exact mechanism unclear. Subtle changes in the microvillus surface have been noted in association with attached EPEC, and this damage may cause diarrhea.

- *Entamoeba histolytica* (causes 10% of liver abscess cases) = noninvasive colitis / fecal leukocyte negative, with bloody diarrhea; 10% of world population infected, however only 10% = clinical disease Infects colon → mimics UC; Treatment = Flagyl [10, 5th ed., pg. 1317]

- Most common causes of diarrhea in AIDS patients = *Cryptosporidium* and *Cytomegalovirus* (CMV) [10, 5th ed, pg. 1319]

- Most common cause of chronic diarrhea in AIDS patients = *Cryptosporidium* or *Isospora belli*

- Most common symptom in AIDS patient = diarrhea

- Most common infection of GI tract in AIDS patient = oral Candida

- *Bacillus cereus* = enteritis after eating fried rice

- Ciguatera fish poisoning = caused by consumption of reef fish that feed on dinoflagellates (algae); most common ciguatoxin carriers: red snapper, grouper, amberjack, sea bass, sturgeon, barracuda; symptoms = N/V/D, paresthesias, *paradoxical temperature reversal*, teeth feel loose, vertigo; treatment = supportive (antihistamines, amitriptyline, fluids); if brady → atropine and dopamine; consider Mannitol in severe cases.

- Scombroid fish poisoning = caused by consumption of dark meat fish (tuna, mackerel, skipjack, bonito, marlin); nonscombroid species (mahi-mahi sardine, yellowtail, herring, and bluefish); histamine-like reaction: flushing, palpitations, HA, N/V/D, diffuse, macular, blanching erythema, *peppery bitter taste*; treatment = antihistamines

- Diarrhea in patients with pet turtle or inguana, asplenic or with sickle cell = **Salmonella**

- Diarrhea after potato salad or mayonnaise = **Staph aureus**

- Diarrhea after eating raw oysters = **Vibrio cholera**

- Most common water-borne diarrhea US = **Giardia lamblia**; Parasite however no Eosinophilia

- Most common symptom with **Giardiasis** = acute watery or pale explosive, offensive smelling diarrhea 90%; abdominal colicky pain / distention /flatulence in 75%. Asymptomatic 15%; [10, 5th ed., pg. 1316]

- Most common intestinal parasite in the US = **Giardia lamblia**; treatment = Metronidazole (Flagyl)

SURGERY / GI / TRAUMA

CAUSES OF POST-OP FEVERS
Mnemonic - (7 W's) [ANK]

W	**W**ind (atelectasis or pneumonia) – first 24 hours
W	**W**ater (UTI) – days 3 to 5
W	**W**anes → veins → check IV sites – days 3 to 5
W	**W**alk (DVT/PE) > 5 days post-op
W	**W**ound > 5 days post-op
W	**W**onder drugs → drug fever
W	**W**omen → endometritis

ARTERIAL OCCLUSION
Mnemonic - (6 P's) [ANK]

P	**P**ain
P	**P**ulselessness
P	**P**allor (limb initially white, with time cyanosis may appear indicating desaturation of blood with ongoing ischemia)
P	**P**olar (for cold, and rhymes with pallor)
P	**P**aresthesias (complete anesthesia = immediate surgical intervention) [10, 5th ed, pg 1191]
P	**P**aralysis (last finding)

- Most common cause of arterial occlusion = embolism
- Embolism originates from the heart (thromboembolism) in 80 to 90% of cases
  Afib and recent MI are the two most common causes of mural thrombus within the heart
  Note: atheroembolism = microemboli from blood vessels
- Most common site of arterial embolism occlusion = bifurcation of common femoral artery
- Ankle Brachial Index: normal > 90%, mild 70-90%, moderate 50-70% and severe arterial insufficiency < 50% (inflate cuff above ankle, doppler → DP or PT, compare to brachial BP)

POST-OP COMPLICATIONS OF THYROIDECTOMY
Mnemonic - (THYROIDS) [3]

T	**T**etany-hypoparathyroidism
H	**H**emorrhage
Y	**Y**ell – recurrent laryngeal → stridor & superior laryngeal nerve → voice fatigue
R	**R**ecurrent hyperthyroidism
O	**O**esophageal damage
I	**I**nfection
D	**D**eaths
S	**S**torm-thyroid storm - tachypnea, tachycardia and fever

SURGERY / GI / TRAUMA

SURGERY / GI PEARLS

- Most common GI disease in the US = gastroenteritis

- Most common cause of vomiting in adults = medications

- Most common cause of upper neuromuscular swallowing dysfunction = CVA
  2nd = polymyositis or dermatomyositis

- Most common cause of lower swallowing dysfunction = intrinsic motility disorder (achalasia, spasm)

- Most common cause of significant LGI bleed = diverticulosis; note 10-15% of BRPR is from UGI source
  [10, 5th ed., pg. 1332]

- Diagnose Bleeding Site: Angiography requires a brisk bleeding rate (0.5-2 mL/min)

  Technetium-Labeled Tagged RBC Nuclear Scan or GI bleeding scan- is more sensitive and can detect bleeding sites of 0.1mL/min

- Most common cause of UGI bleed = PUD; melena present in 33% of LGI bleeds (need 150-200cc of blood in GI tract for minimum of 8 hours for stool to turn black)
  Consider Protonix 80mg IV bolus, then drip of 8mg/hr

- Most common location of gastric ulcer = lesser curvature of body and antrum

- Most common location of stress ulcers = body and fundus; examples = Curling's ulcers (burns); Cushing's ulcers (associated with ICP → head trauma and CNS tumors); sepsis and shock

- Duodenal ulcers > more common than gastric ulcers
  Duodenal ulcers improve with food, gastric ulcers worsen with food

- Gastric ulcers bleed more often than duodenal

- Most common cause of UGI bleeding in pregnancy = esophagitis

- Most common cause of esophageal varices in USA = ETOH abuse

- Most common cause of esophageal varices worldwide = Schistosomiasis

- Start Octreotide on patients with UGI bleed & known esophageal varices for the first 24 hours of hospitalization

  Octreotide (Sandostatin) mimics natural somatostatin (50mcg IV bolus followed by 50mcg/hr drip) → splanchnic vasoconstriction → variceal bleeding (similar to vasopressin without coronary vasoconstriction)

  In addition to Octreotide treat esophageal varices with: PPI [Protonix (Pantoprazole) 80 mg IV bolus, 8 mg/hr drip], IV fluids, blood products (PRBCs, FFP, platelets) and antibiotics
  Erythromycin 250 mg IV x1 before EGD → promotility → improves EGD quality
  Ceftriaxone 1 gm /day x 5 days → ↓ release of endotoxins → ↓systemic vasodilation → prevents coagulopathy and early rebleeding
  TOX Pearl → Octreotide is used in the treatment of refractory hypoglycemia from oral hypoglycemic agents

SURGERY / GI / TRAUMA

- No evidence that NGT placement aggravates hemorrhage from varices or Mallory-Weiss tears [10, 5th ed, pg. 1319]

- Mallory-Weiss tears = partial thickness esophageal tear with bleeding after vomiting

- Boerhaave syndrome = full thickness esophageal rupture after vomiting

- Classic triad presentation of Boerhaave syndrome = vomiting, CP, and subcutaneous emphysema
  SOB may be secondary to pleuritic CP or left-sided pleural effusion (common), or pneumothorax
  Hamman's crunch = Pneumomediastinum; 20% of patients
  Esophagram helps confirm diagnosis; Gastrografin (water-soluble contrast) = 90% sensitivity

- Most common cause of esophageal perforations = Iatrogenic perforations (others:FBs, caustic burns, Boerhaave's)

- Most common site of **Iatrogenic perforations** = pharyngoesophageal junction or the esophagogastric junction [10, 6th edn, pgs 482-485]

- > 90% of **spontaneous esophageal** perforations occur in the distal esophagus

- Foreign Bodies → Most common in kids = 80%

- Most common site for Esophageal foreign bodies to lodge
  #1 = Level of cricopharyngeus muscle (C6)
  #2 = Aortic arch (T4)
  #3 = GE junction

- Most common esophageal impaction in kids = coins

- Most common esophageal impaction in adults = food (meat/bones); tx = glucagon IV, NTG, procardia .... EGD

    Glucagon 1 to 2mgIV
       Relaxes esophageal smooth muscle and the LES
       Little effect on the motility of the proximal esophagus
       Less effective in patients with structural abnormalities, such as strictures or rings
       Success rate 15 to 50%
    NTG, Procardia (may cause hypotension)
    Get EGD!
    Papain (meat tenderizer) – NO! → meat tenderizer may cause necrosis of the esophagus
    If perforation suspected → water-soluble contrast (Gastrograffin)

- CXR esophageal foreign body = frontal plane; tracheal = sagittal plane

- Most common age group to aspirate FB = 1 to 3 years

- Button battery ingestion: if in duodenum → observe; if in esophagus → immediate removal; if not removed in 8 hours there is a risk of esophageal perforation secondary to rapid erosion

- Most common malignancy of esophagus = adenocarcinoma, (no longer squamous cell); adenocarcinoma has continued to rise since 1970 and is now > 50% of all esophageal carcinomas

SURGERY / GI / TRAUMA

- Barrett's esophagus is a risk factor for adenocarcinoma; up to 10% of patients with Barrett's esophagus will develop adenocarcinoma

- Most common benign stomach neoplasms = Polyps (90% hyperplastic/inflammatory polyps, 10% adenomatous – single cauliflower-like malignant transformation risk)

- Most common malignancy of stomach = Adenocarcinoma; Check → CEA, Virchow supraclavicular-node; blacks > whites; lymphoma second most common stomach malignancy

- Krukenberg tumor → Primary stomach or breast CA metastasis to ovary (usually both ovaries; accounts for 5% of ovarian cancers)

- Klatskin tumor = cholangiocarcinoma – cancer of biliary tree

- Most common chronic infection in liver or kidney transplant = CMV

- Most common malignancy of small intestine in the US = Adenocarcinoma

- Most common malignancy of large intestine = Adenocarcinoma

- Most common site of large bowel perforation = cecum

- Most common cause of mesenteric ischemia = arterial embolus; get angiogram not CT
  Other causes = mesenteric arterial thrombosis (15%) or venous thrombosis (15%), non-occlusive mesenteric ischemia (20%) and hypercoagulable states [10, 5th ed., pg. 1288]. *Pain out of proportion to exam*; heme+ stool; ↑lactate

- Toxic megacolon occurs in 5% of case of Ulcerative Colitis; cause unclear; transverse colon more common

- Ulcerative Colitis lesions = erythema nodosum, pyoderma gangrenosum and aphtous stomatitis

- Crohn's = peri-anal disease (fistulas and fissures), bowel malignancies 3x more common

- Most common anorectal abscess = perianal (may be first presentation of Crohn's)

- Most common cause of SBP = *E. coli* 50%, Enterococcus, Strep. Pneumo (Tx = Amp + Gent)

- APT-Downey Test = differentiates if a bloody stool contains maternal or fetal blood → add 1% NaOH to bloody stool → fetal Hgb resists oxidation → remains pinkish-red, whereas maternal Hgb changes to dark brown color [10, 5th ed, pgs. 2307]

TRAUMA PEARLS

- Monocular diplopia = lens dislocation

- Leading cause of traumatic death in adults and PEDS = severe Traumatic Brain Injury; #2 = thoracic trauma

- Severe Traumatic Brain Injury (TBI) = GCS < 8

SURGERY / GI / TRAUMA

- Most common cause of severe TBI = Falls 28% > MVC 20% [CDC, National Center for Injury Prevention and Control; 2006]

- SBP < 90 mmHg, O2 saturation < 90%, PaCO2 < 35 mmHg and T > 100.4 °F correlate → with poor outcomes and are secondary insults to avoid in TBI [26, Vol. 30, No. 21, pg. 258]

- Routine hyperventilation in TBI should be avoided, unless evidence of herniation [J. Neurotrauma 2007;24 Suppl 1:S1-106]

- A single episode of hypotension in patients with traumatic brain injury = double mortality

- Seizure > 20 min after trauma – worse; ↑ possibility of internal injury and development of seizures later

- Seizure Prophylaxis [10, 5th ed., pg 296]
    - Depressed skull fracture
    - Penetrating brain injury
    - Intracranial Hemorrhage
    - Acute SDH or Epidural
    - Prior history of seizures
    - Seizure at time of injury or ED presentation
    - Intubated and paralyzed head injury patient

- Most common traumatic herniation syndrome = Uncus of temporal lobe → transtentorial herniation; CN III compressed → ipsilateral, dilated, non-reactive pupil; contralateral hemiparesis; herniation progresses → decerebrate posturing

- Most common bleed = traumatic SAH

- Most common cause of post-traumatic coma = diffuse axonal injury

- Subdural=crescent shape on CT; bridging veins much more common than epidural and ↑mortality vs epidural

- Epidural = lens / biconvex/ football shaped on CT; middle meningeal artery; "lucid interval"

- Most common CT abnormality after severe closed head injury = traumatic SAH [10, 5th ed., pg. 310]

- Most common bone fractured in children with skull fractures = parietal bone (60-70%)

- 15 to 30% of linear skull fractures in children have been associated with an intracranial injury

- Growing Skull Fractures = linear skull fractures in children that enlarge over time and produce a cranial defect. Result from tear in dura, present months to years following the initial injury, usually require surgical correction

- Indications for Obtaining Head CT in Children with Head Trauma
    - AMS   Focal neurologic deficits
    - HA   Evidence of basilar or depressed skull fractures
    - LOC   Irritability or behavior changes
    - Amnesia   Scalp hematoma in children < 2 years old
    - Seizure   Persistent vomiting

- Significant head or facial trauma have 5 to 10% associated C-spine injuries

SURGERY / GI / TRAUMA

- Nexus Criteria for C-spine imaging (if any criteria present, cannot clear C-spine clinically)
  - Midline spinal tenderness
  - Focal neurological deficit
  - Altered level of consciousness
  - Intoxication
  - Distracting injury present

- Patients with spinal cord injury have 25% associated head injury

- Most common level of C-spine injury in Elderly = C1 to C3 (higher than younger, non-peds patients)

- Most common C-spine fx in Elderly = Type 2 odontoid fx

- Most common cervical fracture in kids = higher cervical (especially odontoid); more common in older kids

- Most common facial fracture = nasal (rule-out septal hematoma); #2 mandible

- Most common Mandible fx = condyle 36% (jaw deviates towards fx on maximal opening) > body 21% angle 20% > symphysis 14% >> ramus 3% [10, 5th ed., pg 326]; most common side = left; similar to pelvic ring usually 2 fractures; subungal or buccal ecchymosis is pathognomonic for mandibular fracture

- **LeForte Fractures** [10, 5th ed., pg. 324]

  - I = **Horizontal** fracture involving only the maxilla at level of nasal fossa
    Motion of hard palate but not nose; if SQ air → sinus fracture

  - II = **Pyramidal** fracture = vertical fxs through maxilla, nasal bones, medial aspects of the orbits
    Motion hard palate and nose, not eyes; blood in nares, rhino or otorrhea; swollen mid-face

  - III = **Craniofacial** disjunction; "dishface"; entire face moves but not head; CSF rhinorrhea

- **Orbital Blow-Out Fracture** = Vertical diplopia - entrapment of inferior rectus muscle → results in *limited upgaze* and may cause pain on attempted upgaze; endophthalmos
  Fractures along the floor usually affect the – infraorbital nerve → hypoesthesia of the cheek and upper

- Weakest area of Orbit = floor (contents prolapse into maxillary sinus)

- **Retrobulbar hematoma** = Eye pain, diplopia, visual loss, reduction of ocular motility, proptosis, ↑IOP, ecchymosis of eyelids, chemosis, ophthalmoplegia, APD; Treatment = lateral canthotomy

- Most common Zygomatic fracture = arch > tripod [12, 6th ed., pg. 1588]; both uncommon

- **Tripod fracture**
  Zygomatico –**Maxillary** articulation - infraorbital rim fx
  Zygomatico - **Temporal** articulation - at the arch
  Zygomatico - **Frontal** articulation

- Flat cheek, diplopia, anesthesia to cheek/upper lip, cheek or periorbital edema = Tripod fracture

- **Zones of the Neck** [10, 5th ed., pg. 371]
  Zone I - (base of neck) extends superiorly from the sternal notch and clavicles → cricoid cartilage
  Zone II - (midneck) cricoid cartilage → angle of the mandible
  Zone III - (upper neck) angle of the mandible → base of skull

- Most commonly injured intra-abdominal structure in PEDS = spleen; Kehr's sign (referred pain shoulder)

- Most commonly solid organ damaged after blunt trauma = spleen

SURGERY / GI / TRAUMA

- Most commonly solid organ damaged after penetrating trauma = liver

- Tension pneumothorax = ↓ BP, distended neck veins, ↓ breath sounds and tracheal deviation

- Flail Chest = > 3 consecutive rib fractures in 2 or more places; produces a free-floating, unstable segment of chest wall → paradoxical chest wall movement

- Major cause of respiratory insufficiency in Flail Chest = pulmonary contusion; Treatment = aggressive pulmonary toilet, intercostal nerve blocks, indwelling epidural catheters, CPAP; judicious fluid administration [10, 5th ed., pgs. 384-385]; ↑ morbidity (pneumonia, sepsis, pneumothorax, ↑admission)

- Most common significant chest injury in PEDS = pulmonary contusion; > 25% often require mech. ventilation

- Indications for thoracotomy = initial chest tube drainage > 20ml/kg or 1,500 ml of blood; unstable VS; > 300-400 cc/hr (or 7cc/kg/hr) for 4 hours; ↑ing hemothorax on CXR; patient decompensates after initial response to resuscitation; use large chest tube (34 – 40F)

- Hemothorax can be seen on upright CXR with 200 to 300cc of blood

- Communicating /open pneumothorax ("sucking chest wound") treatment = EMS = occlusive dressing taped on three sides; ED treatment = chest tube and occlusive dressing; GSW to chest leaves defect (hole) in thoracic wall

- No association of sternal fractures and aortic rupture

- 90% of Blunt Aortic Injuries occur in descending aorta at the isthmus between left subclavian artery and the ligamentum arteriosum; 80% die at scene; 50% who survive die in 24 hours

- Most common area of heart injured in blunt trauma = Right ventricle; most common rhythm = sinus tachycardia; other = A. Fib

- Most common valvulopathy due to chest trauma = Aortic Regurgitation (AR)

- Beck's Triad: 1) Muffled heart tones 2) ↓BP 3) JVD (↑CVP) = cardiac tamponade

- Most common echocardiographic findings with cardiac tamponade = right ventricular diastolic collapse

- Electrical alternans on EKG is pathognomonic for tamponade

- Air Embolism – place patient head and left side down → decrease air leaving through RV outflow tract

- Esophagus is the most rapidly fatal perforation of the GI tract; Hamman's crunch

- Most common site of diaphragmatic injury = Posterolateral; left >> right (liver protects on right); 5% bilateral. FYI: most common site herniated disks rupture, and almost all spinal hematomas = Posterolateral location

- Diaphragmatic injuries are more common in penetrating trauma; CXR diagnostic only 10 to 40%;

- Needle cric if < 8 years old

Indications for Percutaneous Transtracheal Ventilation ("Needle Cric")
1. Cannot control airway with standard interventions or LMA
2. Severe maxillofacial trauma
3. Obstructive processes

Contradintiations for Percutaneous Transtracheal Ventilation ("Needle Cric")

SURGERY / GI / TRAUMA

1. Damaged cricoid cartilage
2. Tracheal rupture
3. Caution with complete upper airway obstruction

Indications for Tube Thoracostomy
- Pneumothorax - Open or closed (moderate to large) or Tension
- Respiratory symptoms regardless of size of pneumothorax
- Hemothorax, Hemopneumothorax, Hydrothorax, Chylothorax, Empyema, large Effusion
- Patients with pneumothorax who are intubated or about to be intubated
- Patients with pneumothorax about to undergo air transport
- Bilateral Ptx regardless of size

Complications of Tube Thoracostomy
pulmonary edema, contralateral ptx, infection (cellulitis, empyema), bronchopleural fistula, pleural leak, formation of hemothorax (lung parenchyma, vs intercostal artery injury), placement (in abdomen, SQ tissue etc), intercostal vessel or nerve injury (avoid inferior margin of rib)

Indications for Emergency Department Thoracotomy (EDT) are determined by: [http://emedicine.medscape.com/article/82584-overview]
- Presence of signs of life
- Mechanism of injury
- Location of injury

Increased thoracotomy survival rates are associated with
- Signs of life in the emergency department (ED)
- Thoracic injuries (as opposed to abdominal injuries)
- Penetrating injuries (as opposed to blunt injuries): Survival in blunt cardiac injury is less than with penetrating cardiac injury secondary to poor cardiac function (due to cardiac contusion) and a higher incidence of associated injuries (cardiac rupture and aortic rupture)
- Stab wounds (as opposed to GSW): GSW injuries are usually unable to spontaneously seal because of large injury pattern. If patients present with any signs of life, they are usually in profound hemodynamic compromise

"Relative" indications for EDT
- Penetrating thoracic injury with traumatic arrest without previously witnessed cardiac activity
- Penetrating nonthoracic injury (eg, abdominal, peripheral) with traumatic arrest with previously witnessed cardiac activity (pre-hospital or in-hospital)
- Blunt thoracic injuries with traumatic arrest with witnessed cardiac activity (pre-hospital or in-hospital)

OB TRAUMA PEARLS

- Primary cause of fetal death in trauma = maternal shock and death
- Second most common cause of fetal death in trauma = placental abruption; 30% of placental abruptions after trauma will not have vaginal bleeding [ATLS, 8th ed., J Trauma 2008;64:1638-1650]
- Placental abruption is a clinical diagnosis = vaginal bleeding, abdominal pain, uterine tenderness, uterine contractions, and fetal distress
- Most common cause of blunt abdominal trauma in pregnancy = MVC 70%; others = falls, direct assault

ORTHOPEDIC

MNEMONICS AND PEARLS HANDBOOK

DESCRIBING ORTHOPEDIC RADIOGRAPHS

Open vs. Closed

Fractures line

Relates to long axis of involved bone (spiral, oblique or transverse)
Simple vs. Comminuted (more than 2 fractures segment)

Location

Which bone is fractured, left vs. right, dominant vs. non-dominant hand, approximate location-proximal, middle or distal 1/3 for long bones, use standard reference points – humeral neck, tibial plateau or intertrochanteric region of femur, extra/intraarticular extension etc.

Position of bone fragments

Displacement: describe the DISTAL fragment in relation to the proximal one (describe in %)
Alignment: describes relationships of longitude axis of one fragment to another
Angulation: any deviation from normal alignment, describe by direction of the apex of the angle formed by the two fragments – give degree and direction (dorsal vs. volar or radial vs ulnar) of deformity. The angle is OPPOSITE the direction of displacement of the distal fragment.

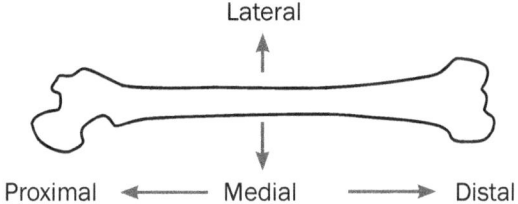

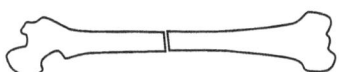
Distraction without displacement or angulation

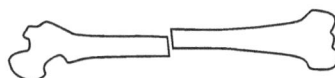
Lateral displacement (25% - 50%) without angulation

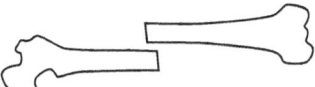
Complete 100% lateral displacement with shortening and without angulation

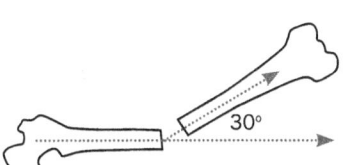
Lateral angulation (30°) without displacement

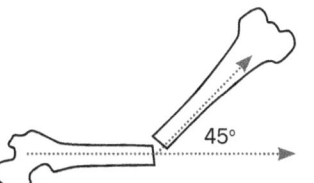
Lateral displacement about 50% and lateral angulation (45°)

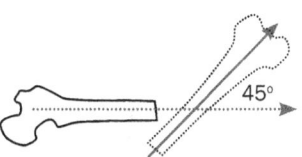
Complete medial displacement with shortening and lateral angulation (about 45°)

ORTHOPEDICS

INTERPRETING C-SPINES
Mnemonic - (ABC's) [ANK/9, with modifications]

A	**Al**ignment
B	**B**ones
C	**C**artilage
s	**S**oft-tissue

See all 7 cervical vertebrae down to the top of T1

Alignment check 3 lines → smooth lordotic curve at 1) anterior and 2) posterior aspect of vertebral bodies and the 3rd line, which is spinolaminar line

- PEDS: 4% of kids < 8 years have pseudosubluxation of C2-C3

Bones check vertebral bodies; ensure anterior and posterior heights are similar
(> 3mm difference suggests fracture)

Cartilage intervertebral joint spaces and facet joints

Soft-tissue
Prevertebral swelling, 6mm at C2 and 22mm at C6
Measure from anterior border of C2 to posterior wall of pharynx (6-at-2 and 22-at-6)
PEDS, < 15 years old, same holds for C2, however at C6 <14mm

Predental space = space from the anterior aspect of the odontoid process and the posterior aspect of the anterior ring of C1; normal predental space =
< 3mm in adults
< 5mm in children
Widen predental space = C1-C2 injury

HISTORY FOR C-SPINE
Mnemonic - (A MUST) [9]

A	**A**ltered mental status

M	**M**echanism
U	**U**nderlying condition
S	**S**ymptoms
T	**T**iming (when symptoms began in relation to event)

Nexus Criteria for C-spine imaging (if any criteria present, cannot clear C-spine clinically)
- Midline spinal tenderness
- Focal neurological deficit
- Altered level of consciousness
- Intoxication
- Distracting injury present

ORTHOPEDICS

UNSTABLE C-SPINE FRACTURES
Mnemonic - (Jefferson Bit Off a Hangman's Thumb) [Dr. Morgan Barnell]

Jefferson	**J**efferson fx
Bit	**B**ilateral facet dislocation
Off a	**O**dontoid fx
Hangman's	**H**angman's fx
Thumb	**T**eardrop fx

POSTERIOR HIP DISLOCATIONS
Mnemonic - (DIP)
Patients who take a "DIP" may have a posterior hip dislocation, most common hip dislocation 80%.

D	A**D**Ducted
I	**I**nternally rotated
P	**P**osterior dislocation (most common hip dislocation)

GALEAZZI'S FRACTURE
Mnemonic - (SURF-surf the sea of Galilee) [11]

S	**S**ubluxated
U	**U**lna (distal radial-ulnar joint (DRUJ)
R	**R**adial shaft (middle/ distal junction)
F	**F**racture

MONTEGGIA FRACTURE
Mnemonic - (BURD) [11]

B	**B**roken
U	**U**lna (proximal)
R	**R**adial head
D	**D**islocation (anterior in 60%)

√ Radial nerve → wrist extension; also → posterior interosseus branch → finger extension; and another branch of radial nerve, which is purely sensory → sensation dorsum of hand [12, 6th ed.,pg. 1691]

- Usually requires surgical fixation

ORTHOPEDICS

COLLES FRACTURE
(Remember Collie = Dog)

D	**D**istal Radial fracture
D	**D**orsal displacement of radial fragment
D	**D**inner-fork deformity

60% with Colles fractures have ulnar styloid fracture

Check median nerve
Reverse Colles = Smith's Fracture
Barton Fracture = dorsal or volar rim fx of distal radius –often fx/dislocations or subluxations

EVALUATION OF ELBOW RADIOGRAPHS IN KIDS
Mnemonic - (**C**areful **R**esurrection **M**edical **T**raining **O**ffers **L**earning)

	Ossification Center	First Appears (year/old) *Note odd numbers*
Careful	**C**apitellum	1
Resurrection	**R**adial head	3
Medical	**M**edical epicondyle	5
Training	**T**rochlea	7
Offers	**O**lecranon	9
Learning	**L**ateral epicondyle	11

Comparison views are helpful for evaluating elbow radiographs in kids
Other mnemonic = CRITOE → I = Internal (Medial) and E = External (Lateral epicondyle)

THE OTTAWA ANKLE RULES

Ankle x-ray series are only required if there is pain in the malleolar zone and any one of the following findings:

1) Bone tenderness along the distal 6 cm of the posterior edge of the fibula or tip of lateral malleolus
2) Bone tenderness along the distal 6 cm of the posterior edge of the tibia or tip of medial malleolus
3) Inability to bear weight both immediately and in the ED for 4 steps [12, 6th ed., pg. 1738]

Foot x-ray series are only required if there is tenderness in the midfoot zone and any one of the following findings:
1) Bone tenderness at the base of the fifth metatarsal
2) Bone tenderness at the navicular bone
3) Inability to bear weight both immediately and in the ED for 4 steps

Ottawa Ankle Rules not developed for patients < 18 years old
Clinical judgment should prevail = if exam is unreliable (ETOH, lack of cooperation, distracting injuries or diminished sensation in the leg) get x-rays

ORTHOPEDICS

BONES OF THE WRIST
Mnemonic - (Some Lovers Try Positions That They Can't Handle) [ANK]

Some	**S**caphoid
Lovers	**L**unate
Try	**T**riquetrum
Positions	**P**isiform
That	**T**rapezium
They	**T**rapezoid
Can't	**C**apitate
Handle	**H**amate

Begin mnemonic:
1st row → RADIAL, proximal row → ULNARLY; scaphoid → pisiform
2nd Row → RADIAL, distal row → ULNARLY; trapezium → hamate

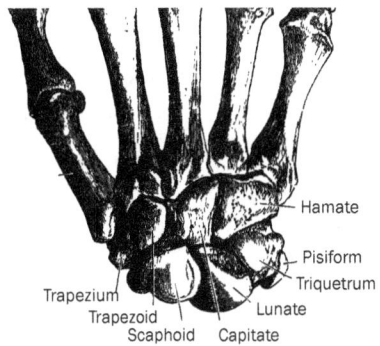

ROTATOR CUFF MUSCLES
Mnemonic - (SITS)

S	**S**upraspinatus → ABDuction; most common rotator cuff muscle injured (85-90%) ("Empty Can Test")
I	**I**nfraspinatus → External rotation
T	**T**eres Minor → External rotation and ADDuction
S	**S**ubscapularis → Internal rotation

All 4 muscles originate on the scapula, transverse the glenohumeral joint and insert on the proximal humerus

FELTY'S SYNDROME
Mneomonic - (FAULTS) [1]

F	Felty syndrome
A	Arthritis – chronic Rheumatoid arthritis (affects < 1% patients with RA)
U	Ulcers of leg (pyoderma gangrenosum)
L	Leukopenia and neutropenia (low PMN's)
T	Thrombocytopenia
S	Splenomegaly (spleen is often FELT)

ORTHOPEDICS

SYNOVIAL FLUID ANALYSIS

Condition	Appearance	WBC's/mm	% PMN's	Glucose; % Serum Level	Crystals Under Polarized Light
Normal	Clear	<200	<25	95-100	None
Non-Inflammatory (i.e. DJD)	Clear	<400	<25	95-100	None
Acute Gout	Turbid	2000-5000	>75	80-100	Negative birefringence; needle-like crystals
Pseudogout	Turbid	5000-50,000	>75	80-1000	Positive birefringence; rhomboid crystals
Septic Arthritis	Purulent/turbid	>50,000	>75	<50	None
Inflammatory (i.e. Rheumatoid arthritis)	Turbid	5000-50,000	50-75	Approx. 75	None

Reference: Clinical Procedures in Emergency Medicine, Roberts et al., 2nd Edition. Harrison's Principles of Internal Medicine, 14th Edition.

ORTHOPEDIC PEARLS

- Most commonly fractured bone in the body = **clavicle** (middle third most common site, 85%); S-shaped bone

- Most common dislocation of shoulder = ANTERIOR; check axillary nerve (pinprick sensation over the skin of deltoid muscle)[12, pg. 1240]; if **shoulder** dislocation secondary to seizure or electric shock than → posterior dislocation (subacromial being common type of posterior dislocations)

- Most common nerve injury with **humeral shaft** fx = radial nerve (wrist and finger extension, sensation dorsum of hand). Other injuries: brachial artery or vein, ulnar and median nerves [10, 5th ed, pg. 563]

- Most common dislocation of **elbow** = POSTERIOR; check ulnar nerve (intrinsic muscles of hand and sensation over ulnar side of hand); vascular injuries 5 to 13%, most common = brachial artery

- Most common bursitis = olecranon bursitis

- Tennis elbow (lateral epicondylitis) confirmed = reproduce pain with
  Elbow in extension, wrist flexion and forearm pronation against resistance [12, pg. 1298] or
  With forearm pronated actively extend fingers or wrist against resistance or
  Pinching with the wrist in extension

- Posterior fat pad on x-ray = supracondylar or radial head fracture (most common elbow fx in adults)

- Most common **carpal fx** = scaphoid > triquetrum 2nd > lunate 3rd

ORTHOPEDICS

MNEMONICS & PEARLS

- Most common dislocated carpal bone = lunate

- Scapholunate gap > 3mm = "Terry Thomas" sign, an indicator of scapholunate dissociation

- Boutonniere Deformity = disruption of extensor hood near the PIP → PIP flexion and DIP extension

- Mallet finger = disruption of extensor tendon at the DIP; immobilize in extension for 8 weeks

- Jersey finger = Avulsion of flexor digitorum profundus tendon (flexed finger is hyperextended); Ring finger most commonly affected [27, Vol. 14, No. 9, pg. 116]

- Boxer's fracture = fx of the 5th metacarpal neck, most common angle in volar direction
  Acceptable Metacarpal Angulation to avoid functional impairment
    - 2nd, 3rd = < 15°
    - 4th = 20°
    - 5th = 40°

- Bennett's fracture = intraarticular base fractures of thumb; unstable; often require surgery

- Rolando's fracture = comminuted Bennett's fracture at the metacarpal base; require surgery; worse prognosis

- Most common cause of **Gamekeeper's thumb** skiing (not twisting the necks of hares); Ulnar collateral ligament (UCL) ruptures 10x more often then radial CL. Treat complete rupture of UCL with surgery, incomplete tear – thumb spicca cast 4 weeks [10, 5th ed., pg. 520]

- UCL rupture if > 35% joint laxity or > 15% more laxity than is present in uninjured thumb; know Stenner's lesion (UCL is prevented from healing by the interposed adductor aponeurosis → chronic instability)

- Finkelstein's test = diagnose De Quervain's tendonitis (tendonitis of the abductor pollicis longus and the extensor pollicis brevis) [12, pg. 1298]

- Phalen's test more sensitive test for carpal tunnel syndrome vs Tinel's sign; nerve conduction studies used to confirm the diagnosis with Sn 90% [10, 5th edn., pgs 547-549]

- Most sensitive bedside test for nerve injury in a finger = two-point discrimination

- Most common dislocation of hip = POSTERIOR; check sciatic nerve and acetabular fx [12, pg. 1255]

- Most common complication with Femoral Neck Fracture = avascular necrosis

- Femoral shaft fracture may result in loss of > 1 liter of blood

- Tibial plateau fx – lateral more common, check peroneal nerve. Cause: direct force driving femoral condyles into articulating surface of the tibia or direct trauma. Check for lipohemarthrosis on x-ray; more than 5mm of depression = surgical repair

- Knee dislocation = ortho emergency = 50% injury to popliteal artery [12, pg. 1261]

- Most common ligamentous knee injury = medical collateral

- Most common dislocation of patella = lateral [10, 5th ed., pg. 688]

ORTHOPEDICS

- Most common cause of acute traumatic knee hemarthrosis = injury to anterior cruciate ligament (ACL)

- Accuracy increases from 70 % using Anterior Drawer test to 99% using **Lachman's Test** in dx ACL injury

- Most common meniscal injury = medical meniscus – less mobile. Knee locks up; Get delayed swelling

- Most common long bone fx = tibia [12, pg. 1263]
- Most common open bone fx = tibia

- Most common site for compartment syndrome in lower extremity = anterior compartment (note: anterior tibial artery deep & peroneal nerve) [12, pg. 1264]; normal pressure = 0 to 8 mmHg; pressures > 30 mmHg cause ischemia

- Most common presenting symptom of compartment syndromes = pain

- Most commonly injured ankle ligament = anterior TALOfibular ligament [12, pg. 1267]

- Maisonneuve fracture = eversion mechanism → proximal fibular fx + disruption of the deltoid ligament or a medial malleolar fx + partial or complete disruption of the syndesmosis

- Pilon fracture = comminuted fracture of the distal tibial metaphysis combined with disruption of the talar dome

- Most common tarsal bone fracture = calcaneus; 75% intraarticular [10, 5th ed., pg. 724]

- Calcaneal fxs rule of 10's = 10% bilateral, 10% vertebral compression fractures, 10% compartment syndrome; 25% lower extremity injuries (tibial plateau fxs, etc)

- Bohler's angle = normal 20 to 40 degrees, if less suspect calcaneal fracture

- Most common midfoot fracture = Navicular (uncommon)

- Most common forefoot fracture = Phalangeal fractures

- Most common metatarsal base fracture = 5th metatarsal

- Most common metatarsal fractured = 3rd metatarsal [10, 5th ed., pg. 725, 728,729]

- **Jones Fracture** = transverse fracture through the base of 5th metatarsal, 1.5cm distal to the proximal part of the metatarsal; treatment = NWB cast for 6 weeks [12, 6th ed., pg. 1745]
  35-50% patients develop persistent nonunions requiring bone grafting and internal fixation
  [http://emedicine.medscape.com/article/825060-treatment]

- **Pseudo-Jones** = avulsion fx, more common, peroneus brevis tendon pulls off a portion of the bone where it inserts; treatment = cast shoe [12, 6th ed., pg. 1745]

- Most common undisplaced metatarsal shaft fracture 2nd-5th tx = below-knee walking cast 2 to 4 weeks
  Non-displaced 1st metatarsal fractures treated with cast 4 to 6 weeks and NWB

- Most common metatarsals involved in stress – "March" – fractures = 2nd and 3rd (fixed)

ORTHOPEDICS

- 2 bones in the hindfoot (calcaneus, talus), 5 bones in the midfoot (navicular, cuboid, 3 cuneiforms), and 19 bones in the forefoot (5 metatarsals, 14 phalanges)

- Hindfoot connects to the midfoot at the Chopart joint; forefoot connects to the midfoot at the Lisfranc joint

- Lisfranc Fracture = disruption of tarsal-metatarsal joint, fracture at base of 2nd metatarsal

- Thompson test check Achilles tendon rupture (plantar flexion weakness/absent) [12, pg. 1303]
  Forced dorsi flexion, palpable gap
  Treatment = open repair or cast immobilization for 8 weeks
  ED management = Gravity Equinus Splint = below-the-knee with ankle plantar flexion

- **Kanavel's 4 cardinal signs** for flexor tendon sheath infection
  1) slight flexion of digit
  2) Swelling ("Sausage Finger")
  3) Tenderness over flexor tendon sheath
  4) Pain on passive extension

- Most common infection with puncture wound to foot = *Pseudomonas*

- Most common cause of osteomyelitis = *Staph aureus*; in sickle cell patients, think Salmonella after staph

- Most common complication associated with leg fractures = infection

- Most common bacteria causing septic joint = *Staph aureus*; most common joint = knee

- Fight bite injury = *Eikenella corrodens*

- Infection of the deep space of the fingertip = felon; treatment = I&D

- Paronychia involves the dorsal aspect of the nail area; treatment = I&D

- Most common viruses causing arthritis = Rubella (German measles) and HBV; others = Parvovirus B19, EBV, mumps, adenovirus and enteroviruses

- Most common joint involved in Gout = Toe; crystals = uric acid; needle shaped, blue; neg. birefringence

- Most common joint involved in Pseudogout = knee; crystals = calcium pyrophosphate; rhomboid, yellow

- Acute Gout/ Pseudogout therapy options = NSAIDs, oral colchicines (more effective with Gout; bad vomiting and diarrhea); cold compresses

- **If contraindication to NSAIDs**
  a) Adrenocorticotropin (ACTH) = Day 1 80 IU IM, q 8h; Day 2 40 IU q 12; Day 3-14 40 IU daily
     → stimulates cortisol → inhibits inflammation [10, 5th ed. Pg. 1592] [Drugs 2008;68:407-415]
  b) Oral prednisone 30 -60 mg daily with taper over 10 to 14 days
     If > 5 joints involved → 3 week taper
  c) Intra-acticular steroid injection – triamcinolone 10mg in knees, 8 mg in smaller joints [Drugs 2008;68:407-415]

- Long term Gout/ Pseudogout therapy = allopurinol (↓uric acid production) or probenecid (↑ uric acid excretion)

ORTHOPEDICS

PSYCHOSOCIAL

RISK FACTORS FOR SUICIDE
Mnemonic - (SAD PERSONS) [ANK]

	Mnemonic	Characteristic	Score
S	Sex	Male	1
A	Age	< 19 or > 45	1
D	Depression	↓concentration, appetite	2

	Mnemonic	Characteristic	Score
P	Previous psych hx	Panic, depression, schizo	1
E	ETOH	Stigmata of ETOH abuse	1
R	Rational thinking loss	Organic brain syndrome	2
S	Separated	Separated > divorced > widow > single > married (least likely)	1
O	Organized attempt	Plan	2
N	No social support	No close family or friends	1
S	Stated future attempts	Determined to repeat	2

Total scores > 6 should undergo psychiatric care directly from the ED

- Females attempt more frequently; males succeed more often
- 20% retry within one year
- 5% of repeaters succeed
- Personality disorders (antisocial, histrionic, narcissistic and paranoid) least likely to commit suicide
- Major Mood Disorder = greatest risk for suicide

DEPRESSION
Mnemonic - (IN SAD CAGES) [ANK]

IN	Interest decrease in everything

S	Sleep disorder
A	Appetite alteration
D	Dysphoric mood

C	Concentration decreases
A	Affect blunted
G	Guilt
E	Energy diminishes
S	Suicide risk

PSYCHOSOCIAL

- 20% of all psych referrals have organic etiologies
  Medical features = abrupt onset, age > 40, visual or tactile hallucinations, abnormal vital signs

- Haldol till they crawl-keep ED safe

- B52 = drop "bomb" on agitated patients = Haldol 5 mg + Ativan 2 mg (can mix the two drugs together in same syringe, "compatible", then give one shot IM

- Most common psychiatric disease = schizophrenia (delusions-most often persecutory, auditory hallucinations, loose associations, catatonia, flat affect)

- Most common cause of dementia in elderly patients = Alzheimer's disease

- Major Mood Disorder = greatest risk for suicide

- The most common DSM-IV diagnostic group for pediatric patients in the ED = substance disorders

  If you have a "difficult patient" – insolvable problems, multiple visits, hostility, name dropping excessive need for attention and threats consider BPD [10, 5th ed., 2605].

BORDERLINE PERSONALITY DISORDER (BPD)
Diagnostic & Statistical Manual (DSM) IV Criteria

A pervasive pattern of instability of interpersonal relationships, self-image, and affects, and marked impulsivity beginning by early adulthood and present in a variety of contexts, as indicated by five (or more) of the following:

1. Frantic efforts to avoid real or imagined abandonment. Note: Do not include suicidal or self-mutilating behavior covered in Criterion 5.

2. A pattern of unstable and intense interpersonal relationships characterized by alternating between extremes of idealization and devaluation

3. Identity disturbance: markedly & persistently unstable self-image or sense of self

4. Impulsivity in at least two areas that are potentially self-damaging (e.g., spending, sex, substance abuse, reckless driving, binge eating). Note: Do not include suicidal or self-mutilating behavior covered in Criterion 5.

5. Recurrent suicidal behavior, gestures, threats, or self-mutilating behavior

6. Affective instability due to a marked reactivity of mood (e.g., intense episodic dysphoria, irritability or anxiety usually lasting a few hours and only rarely more than a few days).

7. Chronic feelings of emptiness

8. Inappropriate, intense anger or difficulty controlling anger (e.g., frequent displays of temper, constant anger or recurrent physical fights)

9. Transient, stress-related paranoid ideation or severe dissociative symptoms

OB-GYN

ECTOPIC PREGNANCY (EP) PEARLS

- 15% of clinically recognized pregnancies terminate in miscarriage

- Mean gestational age for ectopic rupture = 7.0 + 2 weeks

- Recent estimate of heterotopic pregnancy = 1 in 4,000 pregnancies [12, 6th ed., pgs. 721]: if women have undergone embryo transfer or use of ovulation-inducing drugs → incidence = 1 in 100

- Most common site of EP implantation = ampullary (80%) portion of the fallopian tube

- ↑maternal mortality rate if EP implantation = cornual location

- Monoclonal antibody assays detect presence of β-hCG as soon as 2-3 d postimplantation
  The earliest a serum beta-hCG test can detect pregnancy = shortly before missed period
  Usually reaches 200 IU/ml at time of menses [10, 5th ed, pg.2404]

- In a normal pregnancy, β-hCG doubles every 2 days or increase by 66% every 3 days for the first 6-7 weeks beginning 8-9 days after ovulation

- After 9-10 weeks gestation, beta-hCG levels decline

- 10% of normal pregnancies can manifest abnormal doubling times

- 15% of EP's can have normal doubling times

- Evidence of IUP should be seen by transabdominal ultrasound with beta-hCG levels of 6,500 mIU/mL, or at least 1,500 mIU/mL using TVUS (discriminatory threshold)

- Rupture can occur in patients with beta-hCG levels as low as 100 mIU/mL

- The most common clinical presentation of patients with EP = abdominal pain (80%); vaginal bleeding 50-80%) [N Engl J Med 2009;361:379-387]

RISK FACTORS FOR ECTOPIC PREGNANCY

LESSER RISK	GREATER RISK	GREATEST RISK
Previous pelvic or abdominal sx	Previous PID	Pervious ectopic pregnancy
Cigarette smoking	Infertility (IVF)	Pervious tubal surgery or sterilization
Vaginal douching	Multiple sexual partners	Diethylstilbestrol exposure to utero
Age of 1st intercourse < 18 yrs		Documented tubal pathology (scarring)
		Use of IUD

OBSTETRICS AND GYNECOLOGY

ECTOPIC PEARLS

TV Ultrasound findings	beta-hCG (mIU/mL)	Gestational Age
Gestational sac (GS) First sonographic finding	1,000	4-5 weeks
Yolk sac Small hyperechoic ring in GS	1,000 to 7,000	5 weeks
Fetal pole (embryo) Seen adjacent to yolk sac	1,000 to 7,000	5-6 weeks
Fetal cardiac activity Normal rate early preg. = 112 to 136 Slower rates in 2nd and 3rd trimester	10,000 to 23,000	6 weeks

Menstrual age = 2 weeks older than embryonic age (ovulation typically takes place at midpoint of a typical 28-day menstrual cycle) [12, 6th ed., pgs. 717-721]

Double-decidual sac sign ("double ring sign") = Two distinct hyperechoic decidua layers (decidua capsularis and decidua vera) surround the gestational sac (GS). The earliest reliable US finding seen in normal intrauterine pregnancies [12, 6th ed., pg. 717]. Should be visualized by 5 weeks after last menstrual period

Earliest definitive sign in the initial diagnosis of an IUP = **yolk sac** in the uterus; Should be visualized by 5.5 weeks after last menstrual period

Most common cause of misdiagnosis of ectopic pregnancy by TVUS = misinterpretation of the pseudogestational sac as an IUP

Pseudogestational sac = anechoic fluid collection without a clear double decidual reaction
Seen in 10 to 20% of ectopics

Double-decidual sac sign can be used to differentiate true GS from pseudo-GS

Other US findings in Ectopic = free fluid, adnexal mass, or **tubal ring**

Ectopic pregnancies are due to anatomic abnormalities of the salpinx (tube), prior tubal infection or an abnormal endometrium

OB /GYN PEARLS

- Blighted ovum = inability to visualize yolk sac or fetal pole (embryo) in a large gestational sac on TV ultrasound → major criteria for fetal demise

- Subchorionic hematoma = more than ↑double the chance pregnancy loss in threatened abortion [12, 6th edn, pg. 722]

- **HELLP Syndrome** = **H**emolysis, **E**levated **L**iver Enzymes, **L**ow Platelets
  More common in *multigravid* patient
  Can also occur postpartum [12, 5th edn, pg. 697]

- Preeclampsia = HTN, edema, and proteinuria → more common in primigravida patient [12, 5th edn, pg. 697]

- Treatment of Ecclamptic Seizures or prophylactic treatment of patients with severe preeclampsia = Magnesium sulfate first-line agent
  Loading dose 4 to 6 grams over 15 min, followed by IV infusion of 1 to 2 gm/hour maintained for 24 hours after the last seizure

OBSTETRICS AND GYNECOLOGY

Observe for nausea, somnolence;
Serum magnesium level 4 to 8 mEq/L = ↓DTRs
Serum magnesium level > 8 mEq/L = apnea
Treatment of hepermagnesemia = Ca Gluconate [10, 5th edn, pg. 2430; eMedicine]

- Most common cause of death in Toxemia = cerebral hemorrhage [10, 5th edn, pg. 2407]

- Most common medical cause of death in pregnant patient = PE

- Most common cause of death in pregnant patient overall = injury (homicide most common injury)

- The leading cause of death in first trimester is ectopic pregnancy [12, 5th edn, pg. 697]

- Placental abruption usually occurs spontaneously (though trauma can cause abruption) and manifests clinically as painful third trimester vaginal bleeding; clinical diagnosis = dark VB, uterine pain or tenderness is seen in 2/3 and uterine irritability or contractions are seen in 1/3 [12, 5th edn, pg. 697] [10, 5th edn, pg. 2420]; US only 50% accurate in diagnosis. Cocaine is a risk-factor.

- Complications of placental abruption = fetal/maternal death, DIC, amniotic fluid embolism, and fetal/maternal hemorrhage; treatment = delivery

- Placenta previa = common cause of third trimester vaginal bleeding, painless vaginal bleeding occurs. Do not perform vaginal exam if suspected previa/abruption ... it may lead to worsened bleeding

- Most common cause of post-partum hemorrhage = uterine atony

- Most common cause of vaginal bleeding related to primary coagulation disorder = Von Willebrand's disease [12, 5th ed., pgs 673-674 and 1377-1382]

- Most common causes of vaginal bleeding in prepubertal girls (without precocious puberty) = vulvovaginal abnormalities (vaginitis, vaginal FBs, trauma, and tumors)

- Most common organism in lactational mastitis = is S. aureus Treatment = penicillinase-resistant antibiotic and continued emptying of the breast milk (continued nursing or manual extraction). The breast milk will not harm the nursing infant. Women are encouraged to continue to nurse if able. [12, 5th ed., pg 726]

- Uterine size greater than dates = gestational trophoblastic disease and multiple gestation pregnancies

- Molar pregnancy = 1 in 1,700 pregnancies [12, 6th ed., pgs. 677, 723]
    - 80 % present as hydatidiform mole and follow benign course
    - Malignant forms = Invasive Mole (12-15%) and choriocarcinoma (5-8%)
    - Larger then expected uterine size for gestational age and markedly ↑ β-hCG (>100,000 mIU/mL) are risk factors for malignant disease
    - Choriocarcinoma → may metastasize to vagina/lung/liver/brain and is sensitive to chemotherapy

- Hydatidiform mole US findings = intrauterine echogenic mass with multiple small hypoechoic vesicles interspersed = "grape-like" appearance or "snowstorm appearance"

- Hyperemesis Gravidarum = seen in first 12 weeks; occurs in 2% pregnancies; Zofran is ok in pregnancy; abdominal pain is highly unusual and should suggest another diagnosis [12, 5th ed., pg 677]

- Simple cervicitis = ceftriaxone 250 mg IM and a single dose of azithromycin 1gm po

- Chlamydia urethritis = 20% of women with dysuria (sterile pyuria)

- Two most common causes of PID = GC and Chlamydia. Ceftriaxone IM covers GC, the CDC recommends doxycycline for 10 days in PID

OBSTETRICS AND GYNECOLOGY

- Acute complications of PID include
    - Tubo-ovarian abscess (TOA)
    - Peritonitis
    - Peri-hepatitis (Fitz-Hugh-Curtis syndrome)

- Treat asymptomatic bacteriuria in pregnancy = 3-7 days, amox, nitrofurantoin, cephalosporin [Sanford, 2009, pg. 31]

- RhoGAM is indicated for **Rh-negative mothers** who are exposed to a clinical event (miscarriage, ectopic, placenta previa, abruptio placenta, term pregnancy, trauma) that puts them at risk for Rh isoimmunization as this can negatively impact both their current and subsequent pregnancies → hydrops fetalis
    - RhoGAM must be administered within 72 hours of the event
    - 50 mcg IM < 12 weeks gestational age
    - 300 mcg IM > 12 weeks

- Threshold for human teratogenesis = 5-10 rad; fetus most vulnerable 8 to 15 weeks gestation [12, 6th ed., pg. 675]
  1mGy = 0.1 rad

- V/Q — Total fetal exposure to xenon-133 and technetium-99m = 0.5 rad
  CXR = 0.00005 rad; CT head < 0.1 rad, CT chest = < 1 rad; CT abdomen = 3.5 rad [12, 6th ed., pg. 675]

- British Journal of Radiology (2006) [79, 441-444] recommends: D-dimer → + → bilateral lower extremity venous doppler → non-diagnostic → CT chest over V/Q to diagnose PE

- Absolute contraindication to ED speculum and manual pelvic exam = 3rd trimester bleeding

- Tocolytic agents used for the treatment of preterm labor: MgSO4, indomethacin and nifedipine. In the past, terbutaline or ritodrine, were the agents of choice, but in recent years their use has been significantly curtailed due to maternal and fetal side effects = maternal tachycardia, hyperglycemia and palpitations.

- Nitrazine paper pH 7.1 to 7.3 = amniotic fluid; normal vaginal pH in pregnancy = 3.5 to 6.0 [10, 5th ed., pg. 2407]

- Perimortem c-section > 26 weeks gestation, FHTs present; 4 minutes for procedure and 1 min for actual delivery time (70% children who survived were delivered < 5 min) [10, 5th ed., pg. 264]

- Expected physiological changes in pregnancy
    - ↑ HR
    - ↓ SVR
    - ↑ Cardiac Output (HR x SVR)
    - ↑ Blood volume
    - ↓ CVP
    - ↓ SBP which normalizes near term
    - ↑ Minute Ventilation
    - ↑ WBC – mild
    - ↓ BUN / Cr

- 10% of insulin-dependent patients will develop DKA during pregnancy; occurs more rapidly and lower levels of glucose in pregnant patients; hyperemesis and non-compliance most common causes

- pH may be normal in DKA, because the initial pH is ↑ in pregnancy due to physiologic hyperventilation

- **Mondor's Syndrome** = superficial phlebitis of the veins in the SQ tissue of the breast; may occur post-op or minor trauma. (Mondor's disease of penis also described) [10, 5th ed., pg. 1257]

    Menorrhagia = menstruation at regular cycle intervals but with excessive flow and duration
    Metrorrhagia = irregular vaginal bleeding outside normal cycle
    Menometrorrhagia = irregular vaginal bleeding, excessive bleeding, outside normal cycle
    Polymenorrhea = frequent, light, bleeding at intervals < 21 days
    Dysfunctional uterine bleeding = abnormal vaginal bleeding due to anovulataion [12, 6th ed., pg. 648]

OPTHALMOLOGY

ACUTE ANGLE-CLOSURE GLAUCOMA (AACG)

Classic history = patient walks into dark room from daylight or by administering a mydriatic → pupil dilation → occlusion of the chamber angle (the canal of schlemm) → aqueous humor build up → ↑IOP → AACG → periorbital pain, ipsilateral HA, blurry vision / vision loss / halos, abdominal pain

TREATMENT ACUTE ANGLE-CLOSURE GLAUCOMA (AACG)

1) Block aqueous humor production
   a. Topical beta-blocker (Timoptic 0.5% one drop) ↓ IOP in 30 to 60 minutes
   b. Oral /IV Acetazolamide (Diamox) 500mg
   c. Topial alpha-2 agonist – Apraclonidine (Iopidine) one drop

2) Reduce vitreous humor volume
   a. Systemic hyperosmotic agent = Mannitol 1 – 2 gm/kg IV

3) Facilitate aqueous outflow
   a. Pilocarpine 1 or 2% one drop 4x daily → after pressure reduced < 40 mm Hg → makes pupil miotic → pulling the peripheral iris away from the angle
   (will not work if given early because pressure-induced ischemic paralysis of iris)

4) Decrease the inflammatory reaction and reduce optic nerve damage
   a. Topical steroid = Pred Forte 1% one drop every 15 min for four doses then hourly

Other: analgesics, antiemetics and supine position (the lens falls away from the iris decreasing pupillary block)

Differentiate AACG from Iritis = Iritis has normal cornea, constricted to mid-range pupil, normal IOP, ciliary flush (perilimbal injection = dilation of the blood vessels adjacent to cornea), debris in anterior chamber (cell and flare)

HYPHEMA

If hyphema < 1/3 the anterior chamber = manage outpatient – bedrest, **elevate** HOB 30 to 45 degrees, limit eye movement (reading)

Treatment of ↑ IOP = use approach above except the pupil needs to be **dilated** = Atropine 1% one drop three times daily. Atropine will help avoid "papillary play" - constrict/dilate → which stretches iris vessels and promotes bleeding

Avoid Acetazolamide (Diamox) if etiology of hyphema is due to sickle cell disease or if patient is allergic to sulfa

Antifibrinolytic = aminocaproic acid (AMICAR) per ophthalmologist

Major complication = rebleeding after 3 to 5 days; other complications = corneal blood staining, acute/chronic glaucoma, and anterior or posterior synechia formation

OPTHALMOLOGY

CENTRAL RETINAL ARTERY OCCLUSION (CRAO)

- Acute, painless vision loss
- Episodes of Amaurosis fugax
- Pale/gray retina with macular "cherry red spot" (macula is thinnest portion of retina; intact choroidal circulation remains visible through this section of retina)
- Optic disc = boxcar segmentation
- Afferent pupillary defect (APD) (usually not seen with CRVO)
- Causes: embolus (carotid, heart), thrombosis, giant cell arteritis, vasculitis (Lupus), sickle cell disease, hyperviscosity syndromes and trauma [12, 6th ed., 1460-1461]

CENTRAL RETINAL ARTERY OCCLUSION TREATMENT

1. Consult ophtho – determine if anterior chamber paracentesis to lower IOP is indicated
2. Massage 15 seconds with sudden release
3. Topical beta-blocker (Timoptic 0.5% one drop)
4. Oral /IV Acetazolamide (Diamox) 500mg
5. Consider having the patient breath into paper bag for 5-10 minutes if no contraindications → ↑ $PaCO_2$ → vasodilation

Treatment futile if > 90 minutes

CENTRAL RETINAL VEIN OCCLUSION (CRVO)

Dilated tortuous retinal veins, cotton-wool spots, macular edema, and optic disc edema; Retinal hemorrhages, may be mild, moderate or large giving a "blood and thunder appearance"

Ischemic = severe visual loss, extensive retinal hemorrhages and cotton-wool spots, presence of relative APD; complication → neovascular glaucoma

Nonischemic = milder form of the disease, less vision loss, no APD

Causes: HTN, DM, CVdisease, Polycythemia vera, Lymphoma, Leukemia, Clotting disorders, Multiple myeloma, Syphilis, Sarcoidosis, Autoimmune disease – SLE and Oral contraceptive use

No known effective medical treatment is available for either the prevention of or the treatment of CRVO. Identifying and treating any systemic medical problems to reduce further complications is important

ACUTE VISUAL LOSS
Mnemonic - (CAN U GO STARE AT HIM) [http://canadiem.org/2016/01/12/tiny-tip-can-u-go-stare-at-him/ and Dr. Postel]

C	CRVO/CRAO	S	Scleritis	H	Hemorrhage (Hyphema / Vitreous)
A	Abrasion (Corneal)	Ta	Temporal Arteritis	I	Iritis
N	Neuritis (optic)	R	Retinal Detachment	M	Migrane / Meds / Drugs
U	Ulcer (Corneal)	E	Epscleritis		
G	Glaucoma (Acute)	At			
O	Object (Foreign)				

OPTHALMOLOGY

OPTHALMOLOGY PEARLS

- Right eye = OD, left eye = OS. OD acuity appears above the OS acuity when written on the chart

- 20/200 vision OD = right eye sees at 20 feet what a normal eye sees at 200 feet [12, 5th ed., pg 1505]

- Normal intraocular pressure (IOP) = 10 to 20 mm Hg (the # 20 seems important: compartment syndrome suspected if pressures > 20, normal intracranial pressure (ICP) < 20 mmHg)

- Afferent pupillary defect (APD) (Marcus-Gunn pupil) = affected pupil dilates in response to light. Conditions with APD = CRAO and optic neuritis (lesion on retina or optic nerve)

- Most common cause of acute reduction of vision due to optic nerve dysfunction in pt 20 - 40 years of age = optic neuritis; APD; VA reduced rapid and painful vision loss; papilledema; IV steroids = slightly lower 2-year risk of developing MS; oral steroids contraindicated (ONTT = optic neuritis treatment trial) [12, 6th ed., pg 1460]

- Amaurosis fugax = Greek amaurosis = darkening, Latin fugax = fleeting, is a transient monocular visual loss

- Orbital Cellulitis = Proptosis, ophthalmoplegia, edema & erythema of the eyelids, pain on eye movement, fever, headache, and malaise; CT orbits, IV Abx, consult; Up to 11% of cases result in visual loss

- Most common cause of Orbital Cellulitis = ethmoid sinusitis (90%); Risk for cavernous sinus thrombosis

- Hutchinson's sign = HZ lesions on tip nose – prognostic of corneal involvement; herpes zoster ophthalmicus → HZ travels down V1 → nasociliary nerve (branches innervate cornea and skin)

- Retinal detachment = flashing lights, floaters and vision loss. Vision loss: cloudy, irregular, or curtain-like

- Synechia = iris adheres to either the cornea (anterior synechia) or lens (posterior synechia)
  Anterior synechia → prevent drainage of aqueous humor → closed angle glaucoma
  Posterior synechia → block aqueous from posterior chamber to anterior chamber → increased IOP
- Causes = trauma, iritis or iridocyclitis

- Hallmark physical finding in Bechet's Syndrome = hypopyon uveitis is seen rarely; recurrent painful aphtous ulcers of oral mucosa and genitals are more common findings [10, 5th ed., pg 1615]

- Uveitis = uveal tract inflammation; tract consists of three segments, the iris, the ciliary body and the choroid

- Most common cause of blindness in AIDS patient = CMV

- CMV retinitis occurs in 10 to 30% HIV-infected patients [10, 5th ed., pg 1853]

- Treat super glue/crazy glue (cyanoacrylate) = erythromycin ointment [12, 6th ed., pg 1459]

- Alkali burns (liquefactive necrosis) more destructive than acid burns → irrigate copiously, check pH (Normal eye pH = 7.0 to 7.4)

OPTHALMOLOGY

- Cotton wool spots = ↓ retinal blood flow → damage to nerve fibers → puffy white patches on the retina; most common causes = HTN and DM

- Corneal abrasion resulting from tree branch → secondary infection = Pseudomonas aeruginosa

- Corneal abrasion resulting from prolonged contact lens wear → secondary infection = Pseudomonas aeruginosa

ELECTROLYTES

CAUSES OF HYPERKALEMIA
Mnemonic - (RAD MY LAD) [3]

R	**R**enal Insufficiency; RTA type 4
A	**A**drenal Insufficiency
D	**D**rugs K+ sparing = Triamterene and Spironolactone Transcellular shifts = Succinylcholine, β-blockers and digoxin Drug induced hypoaldosteronism = ACE inhibitors, NSAIDs, Heparin, Cyclosporin and Bactrim

MY	**MY**onecrosis / Cell injury (rhabdomyolysis, burns, crush / acute tumor lysis syndrome

L	**L**ack of Insulin → DKA (transcellular shift)
A	**A**cidosis (transcellular shift)
D	**D**igitalis Toxicity (↓K+, ↓Mg2+ and ↑Ca2+→ ↑dig toxicity)

Most common cause of markedly ↑ K+ = lab error [10, 5th ed., pg 1730]

RHABDOMYOLYSIS

- CPK = 5x normal

- Urine = + blood with 0 RBCs

- Most common metabolic abnormality in rhabdo = ↓Ca2+
  Early in rhabdo you get calcium deposition in injured muscle. Later, ↑ Ca2+ secondary to mobilization of deposited calcium and secondary hyperparathyroidism

- Other lab findings: ↑ K+, metabolic acidosis, acute renal failure, DIC

- ↓K+ or ↓Phos = contribute to development of rhabdomyolysis (K+ and phos released from injured muscle so later serum levels may be falsely normal or elevated) [10, 5th ed., pg 1729, 1741]

- Causes: traumatic, exercise induced, toxicologic (CO, toluene, statins, ASA, caffeine, ETOH, neuroleptics/ antipsycotics, cocaine/sympathomimetics), environmental (hypo or hyper-thermia, metabolic (↓K+, ↓Phos, ↑ or ↓Na+, hypo or hyper-thyroidism, DKA or HHS), infectious (influenza most common), snake bites, black widow spider bites, immunologic and inherited

- Treatment: fluid (isotonic crystalloid 500 mL/h and titrate to maintain a urine output of 200 mL/h), urinary alkalinization, mannitol and loop diuretics

- Hemodialysis: persistent hyperkalemia despite therapy, severe acid-base disturbances, refractory pulmonary edema and progressive renal failure.

ELECTROLYTES

TREATMENT OF HYPERKALEMIA
Mnemonic - (C BIG K DROP)[16]

C	**Calcium** chloride or gluconate Calcium gluconate 10cc of a 10% solution slowly over 2 min (stop if bradycardic) Peds = 1.0 mL/kg, not to exceed 10ml, of 10% calcium gluconate solution over 3-5 min If CaCl use central line (CaCl has 3x more Ca2+ than Ca gluconate) Most rapid and effective treatment, stabilizes cardiac membrane without changing serum K+ level Avoid in Digitals toxicity; can use 2gm of Mag sulfate over 5 min if dig-toxic arrhythmia Onset of action = < 5 minutes Duration of action = 30 to 60 minutes Consider repeat dose if EKG changes do not normalize in 3-5 min

B	**Bicarb** (rarely indicated in DKA, unless pH < 7.0 or cardiac arrhythmias) 1 amp Peds Infants = 0.5 mEq/kg IV over 5-10 min Children = 1 mEq/kg IV over 5-10 min In infants use the 4.2% solution and the 8.4% solution in children and adults Onset of action = within minutes Duration of action = 15 to 30 minutes Because of short duration consider Bicarb drip

I	**Insulin** 5-10 units regular insulin IVP (with the dextrose solution) Peds = 0.1 U/kg regular insulin (1 unit regular insulin/ per 5 gram of glucose infused) Onset of action = 20-30 minutes

G	**Glucose** 1-2 amps of D50 Peds = 0.5 gm/kg (2mL/kg) 25% dextrose solution (with insulin over 30 min) Bicarb/insulin/glucose combo "shifts" K+ into cells

Bicarb/insulin/glucose combo "shifts" K+ into cells

K	**Kayexalate** 30-50 gm PO/PR mixed with 100cc of 20% sorbitol Peds = 1 gm/kg/dose PO/PR Resin that exchanges Na+ for K+ in colonic mucosa; Ca/Bicarb/Insulin/Glucose are temporizing measures; kayexalate/diuretics/dialysis = definitive loss of excess K+ Onset of action = 2 to 12 hours Duration of action = 4 to 6 hours

Drop	**Diuretic / Dialysis**

Albuterol 10 mg Neb → ↑ plasma insulin → shifts K+ into intracellular space
Little controversial → tachycardia and chest discomfort
Can be beneficial in patient with renal failure when fluid overload is a concern

ELECTROLYTES

CAUSES OF HYPERCALCEMIA
Mnemonic - (VITAMINS TRAP) [14]

V	**V**itamin D intoxication **V**itamin A intoxication (transcription factor in osteoclast stimulation)
I	**I**mmobilization
T	**T**hyrotoxicosis → direct stimulation of osteoclastic bone resorption
A	**A**drenal insufficiency (Addison's) Lack of hypocalcemic effect of corticosteroid → ↑active Vitamin D → ↑bone resorption & ↑GI Ca absorption; also decrease renal clearance of calcium
M	**M**yeloma / Milk-alkali syndrome → excessive consumption of Ca and absorbable antacids
I	**I**nsufficiency → acute renal failure → ↓Ca2+ ↑Phos → ↑PTH → ↑ Ca2+ from bone and GI and ↓ phosphate reabsorption from kidneys → ↑ phos urine excretion In chronic renal failure the calcium will remain ↓low and ↑phos because kidney cannot excrete
N	**N**eoplasm (squamous cell lung, head and neck, breast, renal, multiple myeloma, leukemia)
S	**S**arcoid (other granulomatous disorders = TB & Wegener's; fungal = Histoplasmosis, Coccidiomycosis) Overproduction of vitamin D by macrophages and ↑extrarenal alpha1-hydroxylase activity (enzyme, which converts Vit D to active form)

T	**T**hiazides calcium carbonate and lithium → alters PTH set-point for inhibition of hormone secretion by circulating Ca
R	**R**habdomyolysis
A	**A**ids
P	↑ **P**TH (parathyroid adenoma 80%) **P**aget disease

Note: HYPER Ca2+ is associated with HYPOkalemia (33% of patients)

90% of cases of Hypercalcemia = malignancy and primary hyper-parathyriodoism

PAGET DISEASE

- Disorder of bone remodeling → excessive bone resorption (osteoclastic activity) followed by a compensatory ↑ bone formation (osteoblastic activity) → structurally disorganized mosaic of bone (woven bone), which is weaker, larger, less compact, more vascular and more susceptible to fracture than normal adult lamellar bone
- Etiology is unknown
- ↑ Ca 2+
- ↑ bone vascularity may → high-output CHF; increased likelihood bleeding complications following surgery
- Most common neurological problem is hearing loss → compression of CN VIII
- Vertebral involvement may lead to nerve-root compressions and cauda equina syndrome

ELECTROLYTES

SIGNS AND SYMPTOMS OF HYPERCALCEMIA
Mnemonic - Stones, Bones, Psychic Moans, Abdominal groans

Stones	renal calculi
Bones	osteolysis = bone pain
Psychic moans	mental status change, seizures, apathy, stupor, coma
Abdominal groans	N/V/anorexia, constipation, PUD, pancreatitis

TREATMENT OF HYPERCALCEMIA

- Initial and most important = rehydration followed by forced diuresis (0.9% NS + Lasix) → ↑ GFR and therefore, excretion in urine. Lasix inhibits resorption of Ca in ascending tubule

- Glucocorticoids - hydrocortisone 200mg/day → inhibits activation of Vitamin D → inhibits bone resorption and GI absorption of calcium

- Plicamycin (mithramycin) → inhibits RNA synthesis in osteoclasts

- Bisphosphates act by inhibiting osteoclastic bone resorption and ↓ viability of osteoclasts (pamidronate and etidronate)

- Calcitonin 4IU/kg SC

- Dialysis

- Correction of underlying condition

HYPOCALCEMIA CAUSES

Renal failure, ↓PTH, ↓Mg, massive transfusions, shock or sepsis, pancreatitis, rhabdomyolysis, Vit D deficiency, alcoholism and drugs

Drugs = cimetidine, phenytoin, phenobarbital, gent, tobra, heparin, protamine, theophylline, nipride, phosphate enemas/laxatives, cisplatin, norepi, loop diuretics, steroids and mag sulfate

HYPOCALCEMIA SIGNS / SYMPTOMS

Depends on serum level and rapidity of decline

Chvostek and Trousseau signs, tetany, seizures, psychosis, ↓ BP, CHF and prolonged QT

Most symptoms = neuromuscular → progressive neuromusc. hyperexcitability [10, 5th ed., pg 1733]

Chvostek sign: a twitch at the corner of the mouth when the examiner taps over the facial nerve (CN VII), just in front of the ear [12, 6th ed., pg. 175]

Trousseau sign: more reliable indicator of ↓Ca = BP cuff maintains a pressure above systolic for 3 minutes → positive if carpal spasm produced [12, 6th ed., pg. 175]

ELECTROLYTES

HYPOCALCEMIA TREATMENT

- Ca Chloride = 360 mg elemental / 14 meq calcium

- Ca Gluconate 90 mg elemental / 4 meq calcium

- 1 amp = 10 cc of 10% solution (either CaCl or Ca Gluconate)

- The ionized calcium will increase for only 1 to 2 hours → follow by repeated doses or an infusion at a rate of 0.5 to 2 mg/kg/hr [10, 5th ed., pg 1733]

ELECTROLYTE PEARLS

- Categories of Hyponatremia
    - Hypervolemic (CHF, cirrhosis, nephrotic syndrome)
    - Euvolemic (SIADH, psychogenic polydipsisa)
    - Hypovolemic (Addison's, renal-GI-third space losses)
    - Pseudohyponatremia (hyperlipidemia, hyperproteinemia)
    - Redistributive (hyperglycemia) – water drawn from cell dilutes Na+

- Correct acute hyponatremia = 1 to 2 mEq/L/hr
  Correct chronic hyponatremia = 0.5 mEq/L/hr

- Required dose of hypertonic saline to correct Hyponatremia =
  (desired Na+ – measured Na+) x 0.6 (weight in kg) = mEq Na+ administered

- Overaggressive correction of hyponatremia = Central Pontine Myelinolysis (CPM) → destruction of myelin in Pons → CN palsies, quadriplegia or coma. More likely to occur in patients with chronic hyponatremia [10, 5th ed., pg 1726]

- Most likely EKG findings in severe hypo-Mg = PVCs and ventricular dysrhythmias (VT, torsades)
  Other EKG findings = AFIB, MAT, PSVT, ↑QT [10, 5th ed., pg 1738]

- Mg is essential cofactor for the **Na+-K+-ATPase pump**
    - Refractory hypo ↓K+ if ↓Magnesium is not corrected along with hypokalemia correction
    - ↓Magnesium → worsen digoxin toxicity induced dysrhythmias

- Most common cause of hyperphosphatemia and hypermagnesemia = renal failure

- 50% of alcoholics = ↓Phos

- Prominent muscle weakness = ↓K+ (paralysis may occur with serum levels < 2.0 mEq/L); also ↓Phos

- Decrease in the serum K+ of 1.0 mEqL = 370 mEq deficit of total potassium (in the absence of acute shifts caused by acid-base disturbances [10, 5th ed., pg 1727]

- Pathophysiology of Hypokalemia from vomiting =
  Alkalosis = K+→ into cells in exchange for H+
  Volume loss → hypovolemia → aldosterone secretion → preserve Na+ & Bicarb in exchange for K+

ELECTROLYTES

ELECTROLYTE PEARLS

- Most important blood protein buffer = hemoglobin
- ↑ or ↓ in pH of 0.10 causes a ↓ or ↑ (opposite change) in PaO2 of about 10% [10, 5th ed., pg 37-38]
  PaO2 = partial pressure of oxygen dissolved in blood
  ↑ acidosis → Hgb gives up O2 more readily → ↑ PaO2 for a particular oxy-Hgb saturation → rightward shift of oxygen-hemoglobin dissociation curve
- pH < 7.35 = Acidosis   pH > 7.45 = Alkalosis
- Normal values: pH 7.35 to 7.45   PaCO2 35 to 45
- PaCO2 and pH move in the opposite direction = Respiratory process
  PaCO2 and pH move in the same direction = Metabolic process
- Acute: for every change of 10 in PaCO2 → the pH moves in the opposite direction (↓ or ↑) by 0.08 + 0.02
- Chronic: for every change of 10 in PaCO2 → the pH moves in the opposite direction (↓ or ↑) by 0.03
- Anion Gap = Na − [HCO3 + Cl]
  Normal range = 5 to 12
- Most common acid-base disorder in seizing patient = respiratory acidosis

CAUSES OF ANION GAP ACIDOSIS
Mnemonic: (CAT MUDPILES) [ANK]

C	CO, CN (inhibit cytochrome oxidase a-a3 → ↑ lactate)
A	Alcoholic ketoacidosis
T	Toluene (secondary) to acidic metabolites

M	Methanol, Metformin
U	Uremia
D	DKA
P	Paraldehyde
I	INH (Isoniazid, inhibits lactate ↔ pyruvate, therefore → ↑ lactate) Iron (hypovolemia and anemia → tissue hypoperfusion → ↑ lactate)
L	Lactic acidosis
E	Ethylene glycol
S	Salicylates

CAUSES OF NON-ANION GAP ACIDOSIS
Mnemonic: (HARD CUP) [ANK]

H	Hyperalimentation	C	Cholestyramine
A	Acetazolamide (Diamox	U	Uterosigmoidostomy
R	RTA (proximal)	P	Pancreatic fistulas
D	Diarrhea		

ENDOCRINE

ADRENAL INSUFFICIENCY (AI)

- Most common cause of Adrenal insufficiency = autoimmune

- Most common infectious cause of Adrenal insufficiency worldwide = TB

- Most common infectious cause of Adrenal insufficiency in US = HIV

- ↓Na+ more common about 90%

- ↑K+ (65%) aldosterone production failure (see hyperkalemia mnemonic)

- ↑ Ca2+ is seen in 6 to 33%

- ↓ Glucose is present in 66% - significant cause of morbidity and mortality

- ↓ BP

- Azotemia from hypovolemia

- ↑hematocrit from hypovolemia; lymphocytosis and eosinophilia (rare)

- Weakness, weight loss, abdominal pain [12, 6th ed, pgs. 1315-1318]

Oral mucous membrane **hyperpigmentation** is pathognomonic primary AI (Addison's Disease) → result of compensatory adrenocorticotropic hormone (ACTH) and melanocyte-stimulating hormone (MSH) secretion

Hyperpigmentation is *not* present in secondary AI (adrenal insufficiency from pituitary infarction or hypothalamic insufficiency)

Treatment in confirmed cases of Adrenal insufficiency = Hydrocortisone 100 mg IV bolus

Treatment in non-confirmed /suspected cases of AI = Dexamthasone [10, 5th ed, pgs. 1782]

Dexamthasone will not affect the serum cortisol level → not interfere with the diagnosis of AI using the cosyntropin stimulation test

Measure cortisol levels → administer cosyntropin (Cortrosyn), a synthetic form of ACTH, → measure serum cortisol levels at 60 min → AI *excluded* if basal or post-stimulation level > 550 nmol/L

ENDOCRINE

NORMAL ADRENAL PHYSIOLOGY

1. Stress → corticotropin-releasing hormone (CRH) (hypothalamus) → Adrenocorticotropic hormone (ACTH) (anterior pituitary) → stimulates the adrenal cortex (zona fasciculata, middle layer of adrenal cortex) → ↑ cortisol

   ACTH is produced from a large precursor protein → in the process, other hormones are generated → ↑ MSH

2. ↓ renal blood flow → renin secretion → angiotensin II production → aldosterone secretion from adrenal cortex (zona glomerulosa /outer layer of adrenal cortex) → ↑ reabsorption of Na+ / water and ↑ secretion of K+ → ↑ BP

   Aldosterone is also produced to a lesser extent by ACTH

   ACTH deficiency doesn't cause mineralocorticoid deficiency, but ACTH excess does cause mineralocorticoid excess

3. Then inner most layer of adrenal cortex = zona reticularis → androgens (DHEA)

4. Adrenal Medulla = core of adrenal gland – catecholamines (epi and norepinephrine)

TREATMENT OF THRYROID STORM

1) Block peripheral effects of thyroid hormone
   a. Propranolol 1 to 2 mg IV q 5 min prn
   b. Guanethidine (inhibits NE release from post-ganglionic adrenergic nerve endings)
   c. Reserpine (depletes stored catecholamines both centrally and peripherally → inhibits release)

2) Inhibit hormone synthesis
   a. Propylthiouracil (PTU) 600 to 1,000 mg po, followed by 250 mg po q 4-6h
   b. Methimazole

3) Block hormone release
   a. Lugol solution or potassium iodide (SSKI) (give one hour after PTU)
   b. Lithium carbonate – difficult to titrate and toxic effects common

4) Prevent peripheral conversion of T4 → T3
   a. Propranolol
   b. PTU
   c. Glucocorticoids (may also be useful in preventing relative adrenal insufficiency due to hyperthyroidism) – Hydrocortisone 100mg IV q 8 hours

5) Provide general support: airway, monitor, fluids, cooling blanket, Tylenol (avoid ASA → ↑ Free T4) [10, 5th ed, pgs. 1772-1774]

Give beta-blocker first = **blocking peripheral adrenergic hyperactivity of thyroid crisis may be the *most* important factor in reducing mortality and morbidity.** If asthma, COPD or CHF substitute for B1 selective medication = esmolol; guanethidine or reserpine = alternatives.

ENDOCRINE

Avoid ASA (see above) and Amiodarone - iodine-rich antidysrhythmic with poorly-defined effects on thyroid function that has been associated with both hyperthyroidism and hypothyroidism

NORMAL PHYSIOLOGY

Thyrotropin-Releasing Hormone (TRH) (hypothalamus) → Thyroid Stimulating Hormone (TSH) or Thyrotropin (anterior pituitary gland) → thyroid gland → capture iodine from blood to synthesize, store & release thyroxine (T4)

Most common cause of hyperthyroidism in US = Graves' disease 80%

MYXEDEMA COMA
Remember - HYPO's

Hyponatremia - common
Hypoglycemia
Hypoventilation → respiratory acidosis
Hypothermia – virtually all cases; Correlates with survival (worst if < 90 degrees F)
↓ HR (Bradycardia)
↓ Hct (Normochromic, normocytic anemia)
↑ Cholesterol > 250 mg/dL
↑ Transaminases, CPK, LDH
Pericarditis → cardiac tamponade
Mental status changes, lethargy, seizures
Erythema nodosum, dry coarse hair, alopecia (lateral 1/3 of eyebrow), hoarse voice, bilateral carpal tunnel syndrome

Myxedema Diagnosis ↓ T4 and ↑TSH

- Most common precipitating factor = infection; others = cold exposure, CVA, CHF, drugs, anesthetics, GI bleed, metabolic disturbances, trauma, surgery

- Amiodarone has been associated with both hyperthyroidism and hypothyroidism

Treatment
Levothyroxine IV + hydrocortisone 300 mg IV followed by 100 mg IV q 6 to 8 hrs

DIABETES INSIPIDUS (DI)

Lose large amounts of dilute urine because of the loss of concentrating ability of distal nephron

Central = lack of ADH secretion from Posterior Pituitary
Nephrogenic = lack of responsiveness to circulating ADH

Causes [10, 5th ed, pgs. 1727]
- Central = idiopathic, infection, tumor, bleed, granulomatous disorders, head trauma
- Nephrogenic = Obstructive uropathy, PKD, renal dysplasia, congenital disorders
- Systemic with renal involvement = Sickle cell, sarcoid, amyloid
- Drugs = Lithium, amphotercin, phenytoin, aminoglycosides

ENDOCRINE

Lab
- ↓ Urine specific gravity
- ↓ Urine osmolality
- ↑ Na+

Treatment of Central DI = parenteral or IN vasopressin

NORMAL PHYSIOLOGY

↑ Plasma osmolarity
↓ BP
↑ Angiotensin II
   ↓
→ Arginine Vasopressin (AVP) or antidiuretichormone (ADH) hormone formed in the hypothalamus → transported via axons to, and released from posterior pituitary → collecting ducts of kidney → reabsorption of water back into the circulation (↑ blood volume)
Vasopressin also → vasoconstrictor (↑ MAP)

ANTERIOR PITUITARY HORMONE RELEASE
Mnemonic: (FLAT PEG) [ANK]

F	Follicle Stimulating Hormone (FSH)
L	Luteinising Hormone (LH)
A	Arenocorticotrophic Hormone ACTH
T	Thyroid Stimulating Hormone (TSH)

P	Prolactin (PRL)
E	Endorphin
G	Growth hormone (GH)

VITAMIN DEFICIENCIES

Thiamine (B1) deficiency
- Alcoholics
- Anorexia, malaise, skin anesthesia, palpitations, calf tenderness, leg heaviness
- Beriberi = edema, JVD, CHF, ↑HR, ↑BP, ↓UO
- Polyneuritis
- Wernicke-Korsakoff Syndrome

Riboflavin (B2) deficiency
- Sore mouth/tongue, stomatitis, glossitis, purple swollen tongue (seen in other B deficiencies)
- Photophobia, loss of VA, corneal ulcers
- Seborrheic dermatitis

ENDOCRINE

Niacin (B3) deficiency - constituent of NAD+ and NADP+
- Pellagra = 3 D's
  - Diarrhea
  - Dermatitis
  - Dementia
- Other: sore tongue, tremors. muscle weakness, anorexia indigestion

Pyridoxine (B6) deficiency
- Peripheral neuritis

Cobalamin (B12) deficiency
- Megaloblastic and pernicious anemias
- Glossitis, hypospermia, GI disorders

Ascorbic Acid (C) deficiency
- Cofactor in collagen synthesis
- Scurvy; swollen/inflamed gums, loosening of teeth, follicular hyperkeratosis
- Impaired wound healing

Vitamin A deficiency
- Night blindness
- Loss of mucous membrane integrity →↑host susceptibility → infection

Vitamin E deficiency
- Peripheral neuropathy, anemia

ENDOCRINE PEARLS

- Most common cause of coma in patient with diabetes = hypoglycemia

- Primary reason for mental status changes in DKA = elevated osmolarity

- Urine dipstick for ketones uses a nitroprusside reaction which measures = acetoacetate ... not beta-hydroxybutyrate; usual ratio in DKA 1:3 acetoacetate / beta-hydroxybutyrate (may be as high as 1:30), therefore urine dip stick does not reflect true level of ketosis

- Average adult fluid deficit in DKA = 5 to 10 liters

- Correct Na+ in DKA = add 1.6 for every 100 mg/dl over the norm

- Diagnostic Features of Hyperosmolar Hyperglycemic State (HHS) per American Diabetic Association

  - Plasma glucose > 600 mg/dL
  - Serum osmolarity > 320 mOsm/kg
  - Serum pH > 7.30
  - Bicarbonate > 15 mEq/L
  - Small ketonuria and absent-to-low ketonemia
  - Profound dehydration up to an average of 9L
  - Some alteration in consciousness

ENDOCRINE

- Most common precipitating factor for Hyperosmolar Hyperglycemic State (HHS) = infection (UTI and pneumonia most common; others = uremia, viral illness, ACS, drugs, metabolic and iatrogenic

- HHS → Start fluid resuscitation first with isotonic saline (0.9% NaCl)
    - Insulin can precipitate vascular collapse if given prior to volume expansion [10, 5th ed., pg 1756]

- Initial treatment of alcoholic ketoacidosis = intravenous D5NS; also, correct K+, add thiamine [10, 5th ed., pg. 2524]

- Most common cause of Hyperparathyroidism = Adenoma 80%; hyperplasia 15-20%, carcinoma < 1%
  ↑PTH → ↑Ca

- The most common presentation of primary hyperparathyroidism = Asymptomatic hypercalcemia

- Pheochromocytoma = catecholamine secreting tumors, usually adrenal; 5 P's = Pressure (BP), palpitations, perspiration, pallor and pain (chest pain, abdominal pain or headache)
  Definitive treatment = surgery

ENVIRONMENTAL

HYPOTHERMIA, HYPERTHERMIA

- Chilblains (Pernio) = inflammatory erythematous to violaceous acral lesions after exposure to cold; pruritic and/or painful; tx = rewarming, gently bandage, elevate; consider Nifedipine (Procardia) [10, 5th ed., pg. 1974]

- Hypothermia = temperature of less than 95°F (35°C)

- Hypothermic patient – move the patient as little as possible – irritable myocardium → VF

- The amplitude of the J-wave (Osborne wave) is proportional to the degree of hypothermia; does not relate to pH and is not prognostic; may appear temp < 32° C (89.6° F) [10, 5th ed., pg. 1981]

- Hypothermia – sequence cardiac deterioration: sinus brady→AF (< 32° C) →VF→Asytole (< 25°C / 77° F)

- Cerebral metabolism ↓ decreases 6% for each 1° C ↓ decline in temperature

- Severe Hypothermia (< 32.2° C, CV instability) → active rewarming → immersion in bath maintained at 40° C (104° F); more practical in ED = Bair Hugger → heat transfer via convection [10, 5th ed., pg. 1991]

  Active core rewarming = peritoneal, bladder and pleural lavage with fluids heated to 45°C (113 °F). From my experience I recommend pleural irrigation by placing two thoracostomy tubes (36 to 40 French) in both hemithoraces

  Lactated Ringer solution is not recommended because hypothermic liver cannot metabolize lactate [http://emedicine.medscape.com/article/770542-treatment]

- DKA /hypothermia → insulin is not effective at core temp < 30° C (86 ° F)

- Defibrillation /hypothermia → rarely effective < 30° C (86 ° F)

- "Core-temperature afterdrop" = further ↓ in core temperature & clinical deterioration after rewarming. Peripheral tissues are warmed → vasodilation → sudden return of cooler, acidotic, hyperkalemic blood from the extremities → central circulation → core-temperature afterdrop → dysrhthmogenic [10, 5th ed., pg. 1975]

- Most common presenting symptom of frostbite = numbness 75% [10, 5th ed., pg. 1973]

- Heat cramps = related to Na+ (salt) deficiency; lab = ↓Na+, ↓Cl , ↓UNa+ and ↓UCl levels; salt your beer!

- Tylenol/ASA not recommended in heat stroke, may be deleterious; tx = evaporative cooling [12, 6th ed., pg. 1188]

ENVIRONMENTAL

SPIDERS, SNAKES, BEES, SCORPION, SEA CREATURES

- **Black Widow bites** = 60% erythematous macule; other findings: target lesions, tiny dual fang marks
  Pathognomonic bite = patch of sweat and a little red dot [Dr. Rangan, ACEP News, Vol. 30, No. 2, Feb 2011]

- **Black Widow Pathophysiology**
  Neurotoxin → ↑ Ach and NE release → SLUDGE BAM + HTN (severe abdominal pain). Tx = supportive, observe 4 hours → if no symptoms discharge home; if HTN and tachycardia do not respond to supportive measures → Equine derived Antivenin available

- **Brown recluse spider** = venom causes local necrotic skin lesion surrounded by an erythematous ring

- **Tarantula** = irritating venomous barbed abdominal hairs, which can be ejected several feet like a javelin → allergic reactions; more painful than damaging (only serious if hairs get into eyes)

- Most common poisonous snake in the US = **Rattlesnake**
  Rattlesnake snakebite = admit patients to the hospital for 24 to 48 hours → serial determinations of platelets, prothrombin time, and urinalysis to check for myoglobin and hemoglobin
  Treatment = Crotalidae Polyvalent Immune Fab (Ovine) (CroFab; FabAV) antivenin - indicated even if envenomation is minimal or mild [http://emedicine.medscape.com/article/771455-treatment]

- **Coral Snakes** admit patients for 24 to 48 hours for observation (delayed signs and symptoms may occur)
  Red on yellow, kill a fellow
  Red on black, venom lack

- *Anyone* bitten by Eastern Coral Snake should be given antivenin IV → 3 to 5 vials in 500 cc NS

- Most consistent symptom after **pit viper** bites = immediate burning pain; petechiae may occur, as well as anaphylaxis if there is an immune response; severe edema, however no compartment syndrome

- Most dangerous venom from **hymenoptera** family = honey bee → venom causes greater histamine release per gram than any other hymenopteran venom [10, 5th ed., pg 793]

- After stinging a victim honey bees release a pheromone that attracts other bees

- **Box Scorpion** = do not treat with narcotic and barbiturates → ↑toxic effects of venom; Antivenin available

- **Jellyfish /man-o-war treatment** = wash off with salt water; 5% acetic acid (vinegar) neutralizes nematocytes

- **Sting ray, starfish, sea urchin, sea cucumbers lion fish treatment** = immersion in hot water [12 6th ed., pg. 1210]

- Hemorrhagic bullous lesions with history of **sea water**-contaminated abrasions or eating raw seafood = consider *Vibrio vulnificus*
  Treatment = Cipro 750 mg po bid [Sanford, 2009, pg. 51]

ENVIRONMENTAL

ELECTRICAL, RADIATION, HAPE/HACE

- Low voltage alternating current (AC) → ventricular fibrillation; High voltage AC → Asystole

- Direct current (DC) → Asystole

- Most common arrest arrhythmia after electrical injury = ventricular fibrillation

- Lightning strike = high voltage DC depolarization; current pathway = "flashover", not horizontal (hand to hand) or vertical (hand to foot) seen with low or high-voltage AC [12, 6th ed., pg 1236]

- "Lichtenberg figures" = superficial ferning pattern = pathognomonic for lightning strike [12, 6th ed., pg 1238]

- Lightning strike = bilateral cataracts and TM perforation common; arrest rhythm = asystole

- Immediate Cause of Death [12, 6th ed., pg 1236]
    - Low voltage alternating current (AC) → ventricular fibrillation
    - High voltage AC → apnea
    - Lightning → apnea

- Absolute lymphocyte count 24 hours after radiation exposure is a good indicator of patients clinical course → > 1,200 → no lethal dose. If lymphocyte count 300-1,200 at 48hrs → lethal dose of radiation expected

- LD50 from exposure to ionizing radiation = 4.5 Gy (450 rad); at 10 Gy (1,000 rad) = 100% MR

- Most sensitive physical finding in HACE = cerebellar ataxia; treatment = descent, O2, dexamethasone

- Most common fatal manifestation of severe high-altitude illness = HAPE; treatment = rest, oxygen, descent, Nifedipine (Procardia, Adalat) 10 to 20 mg po or SL q 6 hours
New study = Dexamethasone and tadalafil (cialis) both decrease systolic pulmonary artery pressure and may reduce the incidence of HAPE in adults with a history of HAPE. Respir Physiol Neurobiol. Dec 15 2011;179(2-3):294-9

- Acetazolamide (Diamox) and dexamethasone have been shown to be effective agents for prophylaxis (not in the treatment) of HAPE

ENVIRONMENTAL

DERMATOLOGY

DERMATOLOGY PEARLS

- **Nikolsky's sign** = Pemphigus vulgaris (autoimmune), TEN and Staph. scalded skin syndrome (the outer epidermis separates easily from the basal layer on exertion of firm sliding manual pressure)

- **Hutchinson's sign** - with HZ infection V1 distribution = vesicle at the tip of the nose

- **Erythema multiforme** = target lesions; 70% mucosal involvement; palms/sole involvement
  Steroids provide symptomatic relief, but unproven benefit in duration and outcome
  Causes: drugs, bugs, immunizations, malignancy (leukemia), idiopathic
  Most common causes: *Mycoplasma*, HSV 1, *Strep. pyogenes* as well as and sulfa, phenytoin and Penicillin drugs

- Most common cause of **impetigo** = *Staph. aureus*; Group A Strep a distant second. Without treatment impetigo heals in 3 to 6 weeks [10, 5th ed., pg. 1639]

- **Tinea versicolor** = yeast infection, Pityrosporum ovale or P. orbiculare are synonyms for Malassezia furfur; variety of colors (tan, pink white), hypopigmented scaly macules or patches on chest or back. Usually seek medical attention because spots do not tan; lesions resolve in 1-2 months without permanent scar

  Recurrence is common [10, 5th ed. pg. 1637]

  Treatement = oral Ketoconazole (Nizoral) 400mg x1 or 200mg q24 hr x 7days or 2% cream 1x q24 hr x 2 weeks; rule-out erythrasma [Sanford, 2009, pg. 104]

- **Erythrasma** = chronic superficial infection intertriginous areas of skin; *Corynebacterium minutissimum*; well-demarcated brown-red macular patches; treatment = Erythromycin 250mg q 6hr x 14 d [Sanford, 2009, pg. 51]

- **Porphyria cutanea tarda** = erosions and bullae to sun-exposed areas exposed to trauma

- **Wood's light** = organism → fluorescent pattern [12, 5th ed., pg. 1572]
  Porphyria cutanea tarda → urine color change to orange or yellow
  Erythrasma → coral red or pink
  Tinea versicolor → green or yellow
  Pseudomonas → yellow or green

- **Pityriasis rosea** = pink or pigmented papules or plaques 1 to 2cm, forming a "Christmas tree-like" distribution on trunk. Usually children and young adults; asymptomatic or pruritis; Eruption is preceded by a week by the appearance of a "herald patch" (2 to 6 cm plaque); self-limiting, resolves in 8 to 12 weeks without treatment; etiology unclear, possibly viral [10, 5th ed. pg. 1638]

- **Poison Ivy** = allergen not in bullae of vesicles; so after washing of the involved site, contact with rash does not cause it to spread; Treatment = antihistamines, oatmeal baths and topical steroids; if widespread oral steroids 30 to 80 mg/day tapered over 21 days [10, 5th ed. pg. 1648] Bentoquatam (IvyBlock) [12, 6th ed. pg. 1249]

DERMATOLOGY

- Steroids in Zoster = treat if within 3 days of onset of rash; ↓ discomfort during acute phase of zoster; does not ↓ incidence of post-herpetic neuralgia [Sanford, 2009, pg 144] or lessen rate of the healing of lesions [10, 5th ed. pg. 1656]

    21 day steroid taper = 30mg bid (days 1-7), 15mg bid (days 8-14) 7.5mg bid (days 15-21) [Sanford, 2009, pg 144]

- **Erythema nodosum** causes = Ulcerative colitis, Yersinia enterocolitica, Srep., Chlamydia, TB, sarcoid, histoplasmosis, coccidiomycosis, hypothyroidism, pregnancy, idiopathic and drugs

- Most common drug-induced cause of erythema nodosum = oral contraceptives

INFECTIOUS DISEASE

SYSTEMIC INFLAMMATORY RESPONSE SYNDROME (SIRS)

More Than 2 Criteria Must be Met

1) Temperature
   a. > 38° C (100.4° F) or
   b. < 36° C (96.8° F)

2) Heart Rate > 90 beats/min

3) Respiratory Rate (respiratory alkalosis is often the first sign of SIRS)
   a. > 20 breaths/min or
   b. PaCO2 < 32 mmHg

4) WBC =
   a. > 12,000 or
   b. < 4,000 cells or
   c. > 10 % bands

Sepsis = SIRS + suspected or documented source of infection

- Severe Sepsis = Sepsis + infection induced organ dysfunction or hypoperfusion
- Organ dysfunction /hypoperfusion = oliguria; lactic acidosis

Septic Shock = Severe Sepsis + refractory ↓ BP despite adequate fluid resuscitation

Sepsis articles
[Crit Care Med. 2008 Jan;36(1):296-327]
[Crit Care Med. 2004 Mar;32(3):858-73]

- Most common source of infection in septic patient = respiratory system [10, 5th ed., pg., 1961]
- Overwhelming Post-Splenectomy Infection (OPSI) = septic shock, DIC and adrenal hemorrhage; may be post surgery or inadequate splenic function (sickle cell) [10, 5th ed., pg 1799]

HIV/AIDS

- Risk of seroconversion after needlestick from HIV + patient = 0.3%

- Risk of HIV from blood transfusion = 1/600,000 to 2,000,000

- Primary HIV occurs 2 to 4 weeks after exposure (Acute Retroviral Syndrome); presentation = mono-like syndrome; Fever, fatigue, sore throat, lymphadenopathy (cervical most common), weight loss, myalgias, HA, N/V/D/, and maculopapular erythematous rash (most commonly on trunk) and leukopenia

INFECTIOUS DISEASE

- Most common opportunistic infection in AIDS patients = Pneumocystis jiroveci (carinii) (PCP)

    - Classic presentation = fever, non-productive cough and DOE which → dyspnea at rest [12, 6th ed., pg. 930]
    - Treatment = Bactrim; if sulfa allergic = clindamycin + primaquine or pentamadine [Sanford, 2009, pg. 41]
    - Steroids if PaO2 < 70 mmHg; steroids 15-30min before meds; reduce resp. failure and death [Sanford, 2009, pg. 41,128]

- Most common cause of serious opportunistic viral infection in HIV patient = CMV

- Most common systemic opportunistic infection in AIDS patient = Mycobacterium avium complex (MAC)
    - Diagnosis AFB in stool or other body fluids
    - Treatment = Rifabutin + Clarithromycin + Ethambutol [10, 5th ed., pg. 1845]

- Most common cause of pneumonia in HIV / AIDS patients = still Strep. pneumo

LYME DISEASE

- Borrelia burgdorferi = spirochete [12, 6th ed., pgs. 970-972]
- Vector = Ixodes dammini / Ixodes scapularis – deer tick
- Zoonotic reservoirs = white-tail deer and white footed mouse
- Most prevalent in Northeast

3 Stages of Lyme Disease Infection

Stage 1	Erythema Chronicum Migrans (ECM) large annular rash with central clearing 2 to 20 days after tick bite
Stage 2	Disseminated phase = 3 days to 6 months after tick bite; fever, adenopathy arthralgias, 50% get multiple annular lesions = most characteristic component of the 2° stage of illness Carditis 8% = 1st degree AVB, Wenkebach, CHB CNS = unilateral or bilateral CN palsies (most common neuro symptom); peripheral neuropathy
Stage 3	Late phase = years after infection = chronic arthritis (knee most common), myocarditis, polyneuropathy or leukoencephalopathy

BABESIOSIS

- **Babesia microti** (NE US) and **Babesia gibsoni** (NW US) and **B. divergens** (Europe) = protozoan
- Vector = Ixodes dammini – deer tick (same vector as Lyme disease)
- Zoonotic reservoirs = deer, rodents and domesticated mammals (cattle, horses, dogs and cats)
- Malaria-like disease; protozoan similar in structure and life-cycle to plasmodia
- Blood transfusions have been implicated in transmission of babesiosis
- 20% have concurrent Lyme disease
- Viral syndrome presentation with ↑ spiking fevers; hepatosplenomegaly; emotional lability
- ↓thrombocytopenia, ↓leukopenia, ↑LFTs, renal failure (dark urine)
- Hemolytic anemia = ↑IBIL, ↑reticulocyte count, ↑LDH, ↓haptoglobin

- Diagnosis = intra-erythrocytic parasite on Giemsa stained peripheral blood smear
    - ("Maltese Cross" formation)

- Treatment = Clindamycin + Quinine or Atovaquone + Zithro [12, 6th ed., pg 973] and [10, 5th ed., pg 1869]

INFECTIOUS DISEASE

RELAPSING FEVER

- *Borrelia recurrentis* = spirochete [12, 6th ed., pg 1256]
- Vector = lice (singular: louse) or ticks
- Zoonotic reservoirs = humans and wild rodents [12, 6th ed. pg 973]
- Viral syndrome presentation
- ↑LFTs, ↓thrombocytopenia, ↓BP; severe cases → meningoencephalitis, DIC, liver failure, myocarditis
- Relapsing fevers
- Diagnosis = thick smear (similar to Malaria)
- Treatment = Tetracycline 200 mg po or Erythromycin 1gm po x one dose

EHRLICHOSIS

- *Ehrlichia chaffeensis; G – Neg coccobacilli*
- Vector = Ixodes
- Zoonotic reservoirs = deer, dogs and other mammals
- Infects circulating leukocytes → fever, viral like syndrome, maculopapular rash 20%
- ↓Leukopenia ↓thrombocytopenia, ↑LFTs; rarely encephalitis and renal failure
- Treatment = Doxy 100mg po bid x 7 to 14 days [12, 6th ed., pg 973]

Q-FEVER

- *Coxiella burnetii*; intracellular, small G Neg
- Vector = tick; more common route of infection = inhalation of organisms from air that contains airborne barnyard dust contaminated by dried placental material, birth fluids, and excreta of infected herd animals
- Zoonotic reservoirs = livestock (cattle, sheep, goats) or cats
- 50% infected with C. burnetii show signs of clinical illness. ↑ Temp (104-105° F), severe HA, myalgias, non-productive cough, N/V/D, abdominal pain, and chest pain. 30-50% of symptomatic infection will develop pneumonia.
- Atypical pneumonia, meningitis, endocarditis, granulomatous hepatitis
- Treatment = Doxy, Azithro or Quinolone

ROCKY MOUNTAIN SPOTTED FEVER

- *Rickettsia rickettsii*; intracellular, Gram Neg
- Vector = Dermacentor tick
- Zoonotic reservoirs = deer, horses, cattle, cats, dogs or rodents
- Triad = fever, HA and rash (begins wrists/ankles spread → trunk/face)
- Treatment = Doxy or Chloramphenicol if allergic

TULAREMIA
- *Francisella tularensis,* small, Gram Neg, aerobic, rod
- Forms of *F. tularensis* infection recognized in humans include: Ulceroglandular (80%, chancer- like ulcer - with raised margins), Glandular, Oculoglandular, Oropharyngeal, Typhoidal/Septicemic and Pneumonic

- Transmission:
    - Direct penetration of the skin (hair follicles, or cuts/abrasions or contaminated by exposure of an infected animal)

INFECTIOUS DISEASE

- Indirectly from bites of deerflies, ticks or mosquitoes (bacterium not isolated in saliva; scratch after bite → introduce infected feces)
- Exposure of mucous membranes with blood or tissue of infected animals
- (rabbits, squirrels, foxes, skunks, mice, rats)
- Ingestion of contaminated food or water
  - Inhalation
    - Regardless of presenting form of tularemia systemic symptoms of *fever with relative bradycardia in 42%*, chills and rigors, myalgias (often prominent in low back), weakness, malaise and headache
    - Treatment: Streptomycin; alternative = Gentamicin

TRAVEL CHEMOPROPHYLAXIS

- http://www.cdc.gov/travel
- Travel Clinic = Robert Citronberg, MD. Traveler's Health and Immunization Center
- Telephone: (847) 663-9500

DENGUE FEVER

- Arbovirus; Most common serious febrile tropical disease after malaria
- Mosquito transmission = *Aedes aegypti* (day biting mosquito)
- Incubation period = 4 to 7 days

- Asymptomatic or viral syndrome presentation with rash:
- Acute onset of severe HA, myalgias and arthralgias *("Break-Bone Fever")*
- Facial flushing, conjunctival injection, retro-orbital pain and facial edema *("Dengue Facies")*
- Rash = macular or maculopapular on trunk spreads → extremities and face

- Note: West Nile virus (transmitted by *Culex* mosquito) = lymphadenopathy → **absent** in Dengue

DENGUE HEMORRHAGIC FEVER (DHF)

- A small percentage of previously infected patients develop → DHF
- Begins as classic Dengue Fever followed by→
- Hemorrhagic pleural effusions, purpura, petechiae, bleeding diathesis

- Diagnosis = ELISA IgM; lab = ↓leukopenia, ↓thrombocytopenia, false ↑Hct, ↑LFTs; ↓Na (most common electrolyte abnormality); ↑PT/PTT, ↓fibrinogen and ↑fibrin degradation products

- Supportive care; Treat fever with Tylenol not NSAIDs due anticoagulant properties. [12, 6th ed., pg 1253]

DENGUE SHOCK SYNDROME

- Circulatory failure (↓BP, Altered mental status, ↑ HR, Altered MS, cool/clammy, narrow pulse pressure with ↑ peripheral vascular resistance

HANTAVIRUSES

- Inhale material contaminated with mouse urine/feces → hemorrhagic fever + renal failure *or*
- Syndrome of severe respiratory failure and shock

INFECTIOUS DISEASE

- ↑leukocytosis with atypical lymphs, ↓thrombocytopenia
- CXR = bilateral interstitial infiltrates in dependent areas
- Death from CV collapse; Mortality rate 6%, if respiratory syndrome mortality much higher [10, 5th ed., pg. 989, 1829]

MALARIA

Most deadly vector-borne disease in the world

- **Plasmodium vivax, ovale, malariae** and **falciparum** = protozoa [12, 6th ed., pgs. 953-958]
- Mosquito transmission = Female Anopheles (bite at dusk and dawn)
- Also direct transmission → blood transfusion & mother → fetus

- Viral syndrome presentation: Paroxysm shivering and chills → followed by ↑fever → when ↓ fever → patient diaphoretic /exhausted → paroxysms of malaria (correspond to length of asexual erythrocytic cycles; merozoites invade RBC's → cells lyse → new merozoites further invade uninfected RBCs); less common symptoms = N/V/D/HA and jaundice

- **P. falciparum** = most deadly form of malaria; complications = cerebral malaria, ↓hypoglycemia (parasites metabolize glucose from RBCs; especially children), metabolic acidosis, severe ↓anemia, renal failure, pulmonary edema, DIC and death [10, 5th ed., pg 1869]

- Blackwater fever = dark urine secondary to RBC hemolysis from high parasitemia

Malaria Diagnosis
- Giemsa or Wright's - stained thin and thick blood smears [10, 5th ed., pg 1869] [12, 6th ed., pgs. 955]

Malaria Treatment
- Plasmodium vivax, ovale, malariae = Chloroquine + Primaquine (covers exoerythrocytic parasites)
- Chloroquine-sensitive P. falciparum → treatment as above
- Chloroquine-resistant P. falciparum →
- Quinine + Doxy (clinda if contraindication, or < 8y/o) +/- Primethamine-sulfadoxine (Fansidar) or
- Mefloquine +/- Doxy or
- Atovaquone-proguanil (Malarone)

Do not use Primaquine if glucose-6-phosphate dehydrogenase deficiency → hemolysis of RBCs

- Quinine given po or IV; if rapid infusion → ↓ profound hypoglycemia [10, 5th ed., pg 1869]
- Other side effects = ↓BP and cardiac dysrhythmias

LEPTOSPIROSIS

World's most widespread zoonotic infection; common in tropical climates

- Leptospira interrogans = spirochete [12, 6th ed., pg 1255-1256] [http://emedicine.medscape.com/article/220563-overview]

- Fresh water contaminated by bovine, pig, canine or rat urine; 2 to 20 days incubation
- Leptospires multiply in the small blood vessel endothelium

- Two syndromes: anicteric (which is self-limiting) 90% of cases and icteric leptospirosis (Weil's disease) which is more severe form characterized by multi-organ failure

INFECTIOUS DISEASE

- Two distinct phases of illness observed in the mild anicteric form → septicemic (acute) phase and the immune (delayed) phase. In icteric leptospirosis, the 2 phases of illness are often continuous and indistinguishable

- Viral syndrome presentation with severe ↑HA, petechial rash which may involve the palate, conjunctival injection, myalgias (calf, low back) and **fever with relative bradycardia**

- → Symptoms resolve after 4 to 7 days, followed by → asymptomatic period or progress directly→more severe disease

- Aseptic meningitis, hepatitis/liver failure, nephritis/RF, uveitis, rash (jaundice and purpura), TTP, HUS, DIC, pneumonitis/consolidation due to alveolar hemorrhage, acalculous cholecystitis, pancreatitis, myocarditis → CHF, Afib and rarely CV collapse

- → May last up to 4 weeks

- Mortality rate in icteric leptospirosis (Weil's disease) = 5 to 40%

- Diagnosis = isolate leptospires from blood or CSF

- Oral Doxycycline or Amoxicillin if treated within first 3 days

- Penicillin or Ampicillin IV for severe cases

LEISHMANIASIS

- Leishmania = intracellular protozoan [12, 6th ed., pg 1258]
- Transmission = Lutzomyia or Phlebotomus → sandflies
- (Rural Africa, Asia, Mediterranean basin, Central/South America, Brazil, India and Sudan
- ↑ Leishmaniasis in returning U.S. military personnel and their dependents from the Middle East, especially from Iraq (mainly cutaneous)

Clinical Syndromes

1) Cutaneous = most common
2) Mucocutaneous = chronic and relentless disease complicated by secondary infections and pneumonia
3) Diffuse Cutaneous = chronic, difficult to treat, few resulting deaths
4) Visceral (Kala-azar or Black fever) = most fatal form caused by Leishmania donovani

- Darkening of skin is characteristic = Kala-azar or black fever; lymphadenopathy
- "Kala-azar" comes from India → Hindi for black fever
- Infiltration of the hematopoietic system → pancytopenia
- ↑ mortality due to secondary infections (pneumonia, TB, dysentery); hemorrhage or severe anemia
- Pentad = fever, weight loss, ↑ hepatosplenomegaly, ↓pancytopenia and hypergammaglobulinemia

- Diagnosis = aspirate bone marrow, spleen, lymph nodes or punch biopsy from ulcer edge
- Stained smears = Leishman-Donovan bodies = stained amastigotes in macrophages
- Treatment of cutaneous leishmaniasis where the potential for mucosal spread is low, topical Paromycin
- Treatment = **Pentavalent Antimonial compounds;** injection only; call CDC for this one
- If failure/resistance, use Amphotercin

INFECTIOUS DISEASE

HOT TUB FOLLICULITIS, FRESH /SEA WATER CELLULITIS, LUDWIG'S ANGINA, NEMATODES

- Hot tub folliculits = **Pseudomonas follicultits** = generally self-limited; no antibiotics [Sanford, 2009, pg. 52]

- Cellulitis after wound in fresh water lake = consider Aeromonas hydrophilia
- Treatment = Quinolone, alternate = Bactrim [Sanford, 2009, pg. 63]

- Ludwig's angina = cellulitis of the soft tissues of the neck and floor of the mouth; life-threatening complication = airway obstruction. Broad-spectrum antibiotics cover gram-positive,
- gram-negative, and anaerobic organisms, ENT consult, CT

- **Nematodes** Treat all 3 with Albendazole or Mebendazole [Sanford, 2009, pg. 52]
    1) *Necator americanus* (hookworm) major cause of anemia worldwide; eosinophilia may be absent
    2) *Enterobius vermicularis* (pinworm) most common parasite infection in the US → anal pruritus
    3) *Ascaris lumbricoides* (round worm) most common nematode infection in the world → Asymptomatic, bowel perforation, cholecystitis, intestinal obstruction, malabsorption, pneumonitis and pancreatitis
       Mortality rate = 5% if complications occur

- **Cystercercosis** = **Taenia solium** (pork tapeworm)
    - New onset seizures
    - CT = calcified lesions
    - Treatment = Mebendazole for acute infection

INFECTIOUS DISEASE

MISCELLANEOUS PEARLS

WMD PEARLS - DERM

CUTANEOUS ANTHRAX

Painless, depressed, black necrotic eschar (anthrax = Greek for "coal")

BUBONIC PLAGUE

- Regional lymph node infection (femoral most common > inguinal > axillary > cervical)
- Erythemetous, warm, very painful, tender, swollen lymph nodes (buboes) with considerable surrounding edema
- Buboes usually nonfluctuant and rarely, suppurate

- Acral gangrene ("black death") = Septicemic plague

- Common Characteristic Features of VHFs & Plague =Petechiae, purpura, ecchymosis, DIC
- Hemoptysis is present in both VHFs and Plague, however other pulmonary findings are uncommon with VHFs

ULCEROGLANDULAR TULAREMIA

- Erythematous, tender papule at the inoculation site → becomes pustular → ulcerates within days
- → The local ulcer is raised with sharply demarcated margins with depressed center (chancriform)

BLISTER AGENTS

- Lewisite
- Nitrogen and sulfur mustards
- Phosgene oxime

- Most sensitive are warm, moist thin areas → perineum, genitalia, axilla, neck & antecubital fossa (thicker skin of hands may be spared)

- On the skin mustard causes no immediate pain sensation (delayed symptoms, hours after exposure)
- Lewisite and phosgene oxime cause immediate pain

- Lewisite antidote = Dimercaprol (British Anti-Lewisite - BAL)

TRICHOTHECENE MYCOTOXINS ("YELLOW RAIN")

- Protein synthesis inhibitors and inhibit mitochondrial respiration and cause bone marrow suppression; 400x more potent than mustard in producing skin injury/ ↑blisters
- Charcoal binds mycotoxins

MISCELLANEOUS PEARLS

SMALLPOX

- Rash begins on face, and upper extremities and all lesions are synchronous, umbilicated, deeply embedded in dermis, painful, involve palm and soles

- Patients are infectious from the time the rash first appears until all scabs fall off

- Smallpox vaccine can lessen the severity/prevent illness if given within 3 days of exposure

WMD PEARLS - PULMONARY
INHALATIONAL ANTHRAX

- Non-specific flu-like illness with non-productive cough; rhinorrhea uncommon
- Patient may improve before acute deterioration within 24 to 48 hours → diaphoresis, dyspnea, stridor, cyanosis, hemorrhagic mediastinitis, septic shock, death; 50% hemorrhagic meningitis
- **CXR**
    - Wide mediastinum, hemorrhagic pleural effusions. Results from U.S. anthrax attacks 2001 from first 10 patients = 7 had infiltrates, multilobar in 3 patients
- **Treatment**
    - Cipro + 1 or 2 additional antimicrobials: Vancomycin, rifampin, PCN, ampicillin, imipenem, clarithromycin, clindamycin or chloramphenicol
    - IV treatment initially before switching to oral antimicrobial therapy
    - Continue oral and IV treatment for 60 days
    - Consider steroids: meningitis, cutaneous or mediastinal edema

PNEUMONIC PLAGUE

- *Productive cough* → Sputum: bloody, watery or less commonly purulent; F /CP / Dyspnea
- CXR = bilateral infiltrates or lobar consolidation; any pattern possible, including ARDS
- Within 24 hours without treatment → fulminant pneumonia associated with hemoptysis, septic shock, DIC, respiratory failure, circulatory collapse and death; 6-10% get meningitis
- **Treatment**
    - Preferred choices = Streptomycin or Gentamicin
    - Alternative choices = Doxycycline or ciprofloxacin or chloramphenicol
    - Treatment for 10 days

PULMONARY TULAREMIA

- Initial picture of systemic illness without prominent signs of respiratory disease: abrupt onset of high fever with relative bradycardia in 42%, chills, rigors, malaise, sore throat, headache and pleuritic CP, myalgias (often prominent in low back) and non-productive cough
- **CXR**
    - Earliest radiographic finding = peribronchial infiltrates
    - Advancing to bronchopneumonia in one or more lobes
    - Hilar lymphadenopathy and effusions are common

- **Treatment - Preferred Choices** [JAMA June 6, 2001-Vol 285, No. 21:2763-2773]
    - Streptomycin – adult: 1 gm IM twice daily; Peds: 15mg/kg IM or IV daily
    - Gentamicin – adult: 5 mg/kg IM or IV daily; Peds. 2.5 mg/kg IM or IV three times daily
    - Treatment for 10 days

MISCELLANEOUS PEARLS

WMD PEARLS

- Inhalational anthrax, pneumonic plague, Q-fever, Ebola = abdominal symptoms

- Inhalational anthrax, bubonic and septicemic plague, pulmonary tularemia, Q-fever pneumonia: Person to person transmission = has not been reported

- Person to person transmission =
  Pneumonic plague, viral hemorrhagic fevers (not yellow-fever), and smallpox

- Decontamination → bleach effective: anthrax, plague, tularemia, Q-fever, cholera, ricin, VHFs, nerve agents, blister agents

- Hypochlorite solution does not inactivate Trichothecene Mycotoxins

BOTULISM

- 7 types of botulism neurotoxins known as types A-G
- exotoxin (botulinus toxin) → descending symmetrical paralysis (GBS: ascending paralysis)
- Pathogenesis
- Inhale / digest → circulation → peripheral nerve synapses → blocks the release of acetylcholine → descending symmetrical paralysis → ptosis, generalized weakness, dizziness, dry mouth, diplopia, blurred vision, dysphonia, dysarthria, dysphagia and respiratory failure
- Treatment = equine antitoxin

CHOLERA

- Enterotoxin →↑ cAMP → secretion of water and chloride ions → massive "rice-water" stool; fluid losses may exceed 5 to 10liters per day
- Treatment
- Fluid and electrolyte replacement
- Ciprofloxacin 1.0 gm po x once OR
- Doxycycline 300 mg po x once
- For children and in pregnancy: erythromycin or trimethoprim-sulfamethoxazole

RICIN

- Inhibits protein synthesis
- Charcoal binds GI exposure of Ricin

BLOOD AGENT

Cyanide
- Is a tissue toxin - the military incorrectly categorizes with blood agent
- Binds cytochrome oxidase and disrupts oxidative phosphorylation →↑ anaerobic metabolism
- HA/N/V/confusion/combativeness/seizures and coma
- Reddish lips (blue in dark skinned patient)
- Odor = **almonds**
- Signs: Initially ↑ HR and BP followed by → ↓ HR and BP and profound metabolic acidosis
- → Respiratory, CNS and myocardial depression (bradycardia → asystole) within minutes of significant exposure

MISCELLANEOUS PEARLS

- **Treatment**
    - Na-Nitrite → converts RBC hemoglobin (Hgb) → to Methemoglobin
    - Methemoglobin → combines with cyanide to form → cyanometh-Hgb
    - Cyanometh-Hgb and free cyanide are detoxified by sulfur transferase (rhodanese) → thiocyanate Thiocyanate is → eliminated in urine

- Rhodanese function ↑s with the availability of sulfur donor
- Na-Thiosulfate is a sulfur-containing compound
- Na-Thiosulfate + cyanometh-Hgb, via rhodanese enzyme → Na-thiocyanate
- Na-thiocyanate → eliminated in urine
- If simultaneous carbon monoxide and cyanide poisoning Na-Thiosulfate should be used ALONE
- Adjunctive therapy: sodium bicarbonate to correct metabolic acidosis & benzodiazepines for szs

NERVE AGENTS

- Sarin, Soman, Tabun, GF & VX
- Inhibitors of acetylcholinestease → cholinergic excess → SLUDGE BAM Syndrome
- Treatment
    - Atropine: large amounts, (10-20 mg), may be needed over 24 hours
    - Pralidoxime chloride (2-PAM, Protopam) reverses the cholinergic nicotinic effects

CHOKING AGENTS

- Phosgene and Chlorine = non-cardiogenic-pulmonary edema
- Delayed symptoms with phosgene; immediate symptoms are noted with chlorine
- Nebulized 3.75% sodium bicarbonate symptomatic improvement to treat chlorine exposures

RSI - PREMEDICATION

DRUG	DOSE	PURPOSE	SIDE EFFECTS
Premedication **LOAD** = **L**idocaine, **O**pioid analgesic, **A**tropine, **D**efasciculating agents			
Lidocaine	1.5 mg/kg	Suppresses cough reflex. May blunt ICP response to intubation	HYPOTENSION Seizure
Fentanyl (Sublimaze)	3-5 μg/kg	Seldom causes hypotension. Analgesia 70x morphine. Give slowly → over 1" because if fast → muscle rigidity	Apnea Seizure
Atropine	0.02 mg/kg	↓ salivation / bradycardia; min dose 0.1mg Brady from (a) laryngoscope stimulating laryngopharynx parasympathetic receptors (b) Sux → direct muscarinic receptor stim	Tachycardia
Vecuronium (Norcuron)	0.01 mg/kg	Pre-paralytic; dose to block fasciculations due to succinylcholine; give 3" before sux	

MISCELLANEOUS PEARLS

RSI - SEDATION

DRUG	DOSE	PURPOSE	SIDE EFFECT
Etomidate	0.3 mg/kg	< 1 min, durn 5 min, ↓ ICP & IOP	N/V, myoclonic excitation
Thiopental (Pentothal)	3-5 mg/kg	Rapid sedn. 30 sec, durn 10 min. ↓ ICP; No analgesia	HYPOTENSION Apnea, asthma
Methohexital (Brevital)	1 mg/kg	Onset < 1 min, durn 5-10min, No analgesia	HYPOTENSION apnea, hiccup, laryngospasm
Midazolam (Versed)	0.1 mg/kg IV or IM	Rapid sedation < 2 min, duration < 30 min No analgesia; + amnesia and muscle relaxation	HYPOTENSION, Apnea ↓ dose if COPD or age > 60
Ketamine (Ketalar)	1-2 mg/kg over 1 min	Status asthmaticus (Bronchodilation) Beneficial in hypotension; Amnesia Analgesia. Onset 1 min, durn 10 min.	HTN, ↑ HR, ↑ICP, ↑ IOP, emergence reactions ↑ secretions
Fentanyl (Sublimaze)	5 µg/kg	Useful in hypotensive patient. Analgesia. Onset 1 min, durn 30 -60 min.	Apnea; Seizure muscle rigidity if given fast

RSI - PARALYSIS

DRUG	DOSE	PURPOSE	SIDE EFFECT
Succinylcholine (Anectine)	1.5 mg/kg 2.0 mg/kg kids	Depolarizing Rapid onset < 1" Short duration < 10" Give IM if no IV access	↓ HR, ↓ BP, Fasciculations ↑ IOP, V+, ↑ intragastric P, ↑ ICP, ↑K+, myoglobinuria,
Vecuronium (Norcuron)	0.1 mg/kg	Non-depolarizing agent Onset 1 to 3 minutes Duration 20 to 60 minutes	Long duration
Pancuronium (Pavulon)	0.1 mg/kg	Non-depolarizing agent Onset 2 minutes Duration 45 minutes	↑ HR, HTN, No CAD
Cisatracurium (Nimbex)	0.4 mg/kg	Non-depolarizing Onset 2 minutes Duration 90 to 120 minutes	Minimal CV effects
Rocuronium (Zemuron)	1.0 mg/kg	Non-depolarizing Onset 1 minute Duration 20 to 60 minutes	Bronchospasm, ↑ HR, Dysrhythmias

MISCELLANEOUS PEARLS

10 P'S OF RAPID SEQUENCE INTUBATION (RSI)

P 1. Perform H & P (MAPLE), consider indications-risks-alternatives

P 2. Preparation: personal-drugs-equipment

P 3. Pulse oximetry-monitor-automated BP device

P 4. Preoxygenate (5 minutes)

P 5. Premedicate: Atropine, Norcuron, Lidocaine or Fentanyl (wait 3 minutes)

P 6. Pressure (cricoid) and continue to assist ventilation with BVM

P 7. Prime (sedation): Etomidate, Thiopental, Versed, Ketamine or Fentanyl

P 8. Paralyze: Succinylcholine, Norcuron or Zemuron

P 9. Placement of ETT

P 10. Post-intubation; verify tube placement and assure adequate sedation for prolonged paralysis

ENDOTRACHEAL TUBE (ETT) SIZE

- # = the inner diameter in mm [12, pg. 40]
- ETT size adult males = 8.0-8.5 advance tube to 23cm (from carina to corner of mouth) [12]
- ETT size adult females = 7.5-8.0 advance tube to 21 cm (from carina to corner of mouth) [12]
- After intubation and NGT get PCXR, ETT should be 2 cm above the carina; check end-tidal CO2

PRE-INTUBATION ASSESSMENT FOR DIFFICULT AIRWAY
Mnemonic: (LEMON)

L	Look externally
E	Evaluate (3:3:2 rule) = 3 fingers between incisors; mandible length 3 fingers from tip of chin to Hyoid bone and distance of the hyoid to the thyroid – 2 fingers distance
M	Mallampati classification (I-IV); I fully visible tonsils; IV only hard palate visible)
O	Obstruction
N	Neck

MISCELLANEOUS PEARLS

POST-INTUBATION PROBLEMS
Mnemonic: (DOPE)

D	**D**islodged ETT
O	**O**bstructed ETT
P	**P**neumothorax
E	**E**quipment failure

KETAMINE PEARLS

- Do NOT give Ketamine: < 6 months, weight < 5 kg, active pulmonary infection, PMH of cardiac-thyroid-psych or porphyria.

- Note: recent studies = ok to use Ketamine in patients with traumatic brain injury (TBI) and intracranial hypertension → hemodynamic stimulation → improves CPP → prevents secondary penumbra ischemia
[http://emedicine.medscape.com/article/80222-treatment; Neurosurg Pediatr. 2009 Jul;4(1):40-46]

- In conscious sedation may give 4-5 mg/kg IM onset 5 minutes, duration 30 minutes
- Consider atropine (decreases secretions), however not routinely recommended anymore [Acad Emerg Med. 2008 Apr;15(14):314-318]
- Consider versed, IV or IM = 0.1 mg/kg, PR 0.75 mg/kg onset 15" durn 45" - drug of choice for emergence reaction

PEDIATRIC AIRWAY PEARLS

ETT size can be estimated:

a) 16 + (age in years) / 4
b) Age / 4 + 4
c) < 2 y/o = size of patients little finger (fifth digit)
d) Broslow Tape

- Position of the ET tube at the lips (in cm's) should = 3 x size of ETT
- ETT x 2 = NG and Urinary Catheter size

PEDIATRIC TUBE SIZES

	Neonate	6 months	1 to 2 years	5 years	8 to 10 years
Chest tube	12-18	14-20	14-24	20-32	28-38
NGT	5-8	8	10	10-12	14-18
Urinary Catheter	5-8 (feeding)	8	10	10-12	12

MISCELLANEOUS PEARLS

PEDIATRIC AVERAGE WEIGHTS AND ENDOTRACHEAL TUBE SIZES

Age	Weight range (kg)	Endotracheal Tube Sizes)
Premature	1.0-2.5	2.5-3.0
Newborn– 3months	2.5-6.0	3.0-3.5
4-18 months	6-12	4.0-4.5
1.5-3 years	12-15	4.0-4.5
3-5 years	15-20	4.5-5.0
5-7 years	20-25	5.5-6.0
8-10 years	25-35	6.0 cuffed
11-12 years	35-40	7.0 cuffed

[Reference for above chart: "Tube Sizes" and "Average Weights – Endotracheal Tube Sizes" = Larry B. Mellick, MD]

Weight in kilograms = (Age x 2) + 8
Upper limit of SBP = (Age x 2) + 80

Newborn = 3 kg
1 y/o = 10 kg
5 y/o = 20 kg
10 y/o = 30 kg

DRUG INFUSIONS

DRUG	DOSAGE RANGE	INFUSION RATE
DOPAMINE Conc: 100mg/5cc 60 mg/100 cc D5W	2-30 mcg/kg/minute Starting dose: 3 mcg/kg/minute	1 cc/hr = 10 mcg/minute
DOBUTAMINE Conc: 250 mg/20 cc 60 mg/100 cc D5W	2-30 mcg/kg/minute Starting dose: 3 mcg/kg/minute	1 cc/hr = 10 mcg/minute
PROSTAGLANDIN Conc: 0.5 mg/cc Premix: 0.5 mg/83 cc D5W	0.05-0.2 mcg/kg/minute Starting dose: 0.05 mcg/kg/minute	1 cc/hr = 0.1 mcg/minute
ISUPREL Conc: 1 mg/5cc 3 mg/250 cc D5W (add 1 mg to premix)	0.05-1.0 mcg/kg/minute Starting dose: 0.05 mcg/kg/minute	1 cc/hr = 0.2 mcg/minute
EPINEPHRINE Dilution: 1:10,000 Conc: 1 mg/10 cc Premix: 3mg/250 cc D5W	0.05-1.0 mcg/kg/minute Starting dose: 0.05 mcg/kg/minute	1 cc/hr = 0.2 mcg/minute
NOREPINEPHRINE Conc: 1 mg/cc Premix: 3 mg/250 cc D5W	0.05-1.0 mcg/kg/minute Starting dose: 0.05 mcg/kg/minute	1 cc/hr = 0.2 mcg/minute

MISCELLANEOUS PEARLS

LIDOCAINE Conc: 100 mg/5cc Premix: 2 gm/250 cc D5W	Starting dose: 20-50 mcg/kg/minute IV Rate: Run at 0.25-0.6 cc/kg/hr = 20-50 mcg/kg/minute	1.5 cc/hr = 200 mcg/minute
NIPRIDE Conc: 50 mg/5 cc Premix: 50 mg/250 cc D5W	0.5-8 mcg/kg/minute Starting Dose: 0.5 mcg/kg/minute	1.5 cc/hr = 5 mcg/minute
NITROGLYCERINE Premix: 50 mg/250 cc	0.5-20 mcg/kg/minute Starting dose: 0.5 mcg/kg/minute	1.5 cc/hr = 5 mcg/minute
CA CHLORIDE Premix: 10 gm/100 cc	10 gram/100 cc = 100 mg /cc Starting dose: 10-20 mg/kg/hr	1 cc/hr = 100 mg/ hour

Anaphylactic shock: Give Epinephrine IV over 10 minutes

Epi 1:1,000 ---> take 0.1 cc = 100 µg (0.1mg) ---> dilute in 10 cc NS = 100,000 concentration
Or
Epi off crash cart = 1:10,000 (1mg) ---> take 1cc = 100 µg (0.1mg) ---> dilute in 10cc of NS to get 1:100,000 concentration ---> give slowly over 10 minutes

LOCAL ANESTHESIA PEARLS

- 2 "i's" in generic denotes an amide
  Examples: Lidocaine, Bupivacaine, Prilocaine, Mepivacaine and Etidocaine

- If allergic to amides (see above) and esthers, for example procaine (novocaine) →
  consider benadryl → dose = 1cc (50mg/ml) diluted in 9cc of NS

- Maximum dose of lidocaine of infiltration →
  Without epinephrine = 5 mg/kg
  With epinephrine = 7 mg/kg

- Avoid TAC (tetracaine, adrenaline, cocaine) or LET (lidocaine, epinephrine and tetracaine) on mucosal membranes, pinna of the ear, nose, penis, fingers and toes [10, 5th ed. pg., 2574]

- XAP = lidocaine, adrenaline (epinephrine) and pontocaine (tetracaine)
- LET = lidocaine, epinephrine and tetracaine

- Bupivacaine is not recommended for use in children < 12 y/o [10, 5th ed. pg., 2573]

- Consider 6-0 fast-absorbing catgut sutures with facial lacerations [Dr. Robert Rifenburg]

MISCELLANEOUS PEARLS

THE #20 - AN IMPORTANT NUMBER

- Normal IOP < 20 mm Hg
- Normal intracranial pressure (ICP) < 20
- Compartment syndrome suspected if pressures > 20 mm Hg
- If pressure > 30 mmHg → ischemia and pain
- Peak methanol levels > 20 mg/dL = Indication for dialysis [12, 6th ed., pg. 1069,1070]
- Ethylene glycol levels > 20 mg/dL = Indication for dialysis [12, 6th ed., pg. 1069,1070]

POSTEROLATERAL - AN IMPORTANT WORD

- Most common site of diaphragmatic injury = Posterolateral
- Most common site herniated disks rupture = Posterolateral
- Almost all spinal hematomas = Posterolateral
- Most common site of esophageal tear in Boerhaave's syndrome = left Posterolateral aspect (distal esophagus)

REFERENCES

[ANK]   Author is Not Known

[1]   Rogers PT: The Medical Student's Guide to Top Board Scores. Chicago: Innovative Publishing and Graphics, Inc. 1992; pgs. 116, 129, 131, 189, 193, 195, 200, 206, 225.

[2]   Dr. Antonio Carlino

[3]   Dr. Karen Spurgash

[4]   Dr. Tajudeen Ogbara

[5]   Author = Dr. Eric Farinas, mnemonic provided by Dr. Nancy Bauer.

[6]   Wilson SE, et al: Current Clinical Strategies-Surgery 2n ed. Fountain Valley, CA: CCS Publish. 1995:pg. 74

[7]   Adler J, Plantz SH: Emergency Medicine Pearls of Wisdom 3rd ed. St. Louis: Mosby-Year Book, Inc. 1994.

[8]   Armstrong P, Wastie ML: Diagnostic Imaging 2nd ed. Chicago: Year Book Medical Publishers. 1987:pg. 106.

[9]   Markovchick VJ, et al: Emergency Medicine Secrets. Philadelphia: Hanley & Belfus, Inc. 1999: pgs. 63, 199, 355, 408.

[10]   Rosen P, Barkin E, et al: Emergency Medicine Concepts and Clinical Practice. Chicago: Mosby Year Book. 5th edition 2002, 6th edition 2006 and 7th edition 2010.

[11]   Mnemonic provided by Dr. J. Walters from Dr. P. Thomas Rogers.

[12]   Tintinalli JE, et al: Emergency Medicine: A Comprehensive Study Guide. St. Louis: McGraw-Hill 1996 5th edition 2000, 6th edition 2004 and 7th edition 2011.

[13]   ANK/Author is Not Known-mnemonic provided by Dr. John Carroll.

[14]   Dr. Dane Nichols

[15]   Circulation. 2005;112:e28-e32 Management of Massive Pulmonary Embolism

[16]   Dr. Thomas R. Scaggs

[17]   Bhushan E, et al: 1993 First Aid for the Boards 3rd edition. New Jersey: Prentice Hall. 1993; pg. 74

[18]   Author = Dr. Gueyikian, mnemonic provided by Dr. Suzanne Ahn

[19]   Dr. Nancy Bauer

REFERENCES

[20] Dr. George Hevesy

[21] Blackbourne LH: Surgical Recall 2nd edition. Philadelphia: Williams & Wilkins. 1998 pg. 138

[22] ACC/AHA Guidelines for the management of patients with acute myocardial infarction: executive summary and recommendations. Circulation 1999; 100:1016-1030./ AHA 2005 guidelines/ AHA 2010 guidelines

[23] Bosker G: Textbook of Adult and Pediatric Emergency Medicine 1st edition. Atlanta: American Health Consultants, 2000: pgs. 567, 578, 579.

[24] Dr. Robert Rifenburg

[25] www.emedhome.com

[26] Adult EM Reports

[27] Pediatric EM Reports

[28] http://content.onlinejacc.org/cgi/content/full/50/7/652 (ACC/AHA 2007 Guidelines UA/NSTEMI)

[29] 2010 AHA Handbook of Emergency Cardiovascular Care for Healthcare Providers

REFERENCES

REFERENCES

www.ingramcontent.com/pod-product-compliance
Lightning Source LLC
Chambersburg PA
CBHW080909170526
45158CB00008B/2052